F. C. M. Lau · C. K. Tse

Chaos-Based Digital Communication Systems

Springer-Verlag Berlin Heidelberg GmbH

Engineering ONLINE LIBRARY

http://www.springer.de/engine/

F. C. M. Lau · C. K. Tse

Chaos-Based Digital Communication Systems

Operating Principles, Analysis Methods, and Performance Evaluation

With 93 Figures

Springer

F. C. M. Lau
C. K. Tse
Department of Electronic
and Information Engineering
Hong Kong Polytechnic University
Hong Kong
China

E-mails: encmlau@polyu.edu.hk
encktse@polyu.edu.hk

Cataloging-in-Publication Data applied for
Bibliographic information published by Die Deutsche Bibliothek.
Die Deutsche Bibliothek lists this publication in the Deutsche Nationalbibliografie;
detailed bibliographic data is available in the Internet at <http://dnb.ddb.de>

ISBN 978-3-642-05616-1 ISBN 978-3-662-05183-2 (eBook)
DOI 10.1007/978-3-662-05183-2
http://www.springer.de

Originally published by Springer-Verlag Berlin Heidelberg New York in 2003.
Softcover reprint of the hardcover 1st edition 2003

Typesetting: Digital data supplied by authors
Cover-Design: design & production GmbH, Heidelberg
Printed on acid-free paper 62/3020 hu 5 4 3 2 1 0

To our families

Karen, Valerie and Michelle

Belinda and Eugene

Preface

In the 1970's and 1980's, we saw phenomenal advancement in nonlinear science, which had led to many important discoveries that greatly improve our understanding of the physical world. Among them, the discovery of *chaos* in deterministic systems is unarguably one of the most revolutionary scientific findings. We are now able to explain the apparent *complexity* and subtle *order* exhibited by many physical systems under the unified framework of chaos theory.

The past decade has seen heightened interest in the exploitation of chaos for useful applications in engineering systems. One application area that has attracted a great deal of attention is communications. Chaotic signals, by virtue of their wideband characteristic, are natural candidates for carrying information in a spread-spectrum communication environment. The use of chaotic signals in communications thus naturally inherits the advantages that are currently being offered by conventional spread-spectrum communication systems, such as robustness in multipath environments, resistance to jamming, low probability of interception, etc. In addition, chaotic signals are easy to generate and hence offer a potentially low-cost solution to spread-spectrum communications.

Although many practical problems need to be solved before chaos-based communications can be realized in practice, the field has advanced rapidly during the past few years and it now reaches a point where abstract concepts from physics and mathematics have been fruitfully ported to techniques that allow information to be carried by chaotic signals.

This book is intended to address the basic system design, operation, analysis, and performance evaluation of a few selected chaos-based digital communication systems. We put emphasis on the analytical approach taken to study chaos-based communication systems, and focus our attention on a few performance aspects that are of practical importance. In particular, we discuss in this book the modulation techniques, error rate calculations, anti-jamming capabilities, and coexistence with conventional communication systems. We believe that the materials covered in this book will be useful to graduate students, researchers, communication engineers and technology developers who wish to exploit chaos for communication applications.

We begin in Chapter 1 with an overview of spread-spectrum communication systems and the potential benefits of carrying information with chaotic signals. In Chapter 2 we introduce chaos-based digital modulations and discuss the salient concepts in encoding information with chaotic signals. A review of the various modulation schemes is given. Chapter 2 also contains a discussion on the use of equivalent discrete-time baseband models for studying chaos-based digital communication systems. This kind of models will be used throughout the book. Chapters 3 and 4 provide an indepth treatment of analytical techniques for calculating bit error probabilities of chaos-based digital communication systems. Both single-user and multi-user systems are studied. Being spread-spectrum systems, chaos-based communication systems should inherently resist jamming or be designed to do so. Chapters 5 and 6 are devoted to the study of anti-jamming capabilities of chaos-based digital communication systems under narrowband tone jamming and wideband pulsed-noise jamming. Furthermore, if chaos-based systems would represent an alternative or competing technology for future communication systems, a transition can be envisioned when chaos-based systems are put in real practice, and such a transition will inevitably see the coexistence of conventional systems and chaos-based systems. In Chapters 7 and 8, the issue of coexistence of conventional communication systems and chaos-based systems are addressed. Finally, the technologies that enable chaos-based communication systems are still immature, and there is much room for basic research. In Chapter 9, some non-coherent non-correlator-type systems are presented, the aim being to arouse further thoughts in the exploitation of chaos for digital communications.

For the successful completion of this book, we are indebted to several people, institutions and organizations. First of all, we acknowledge with gratitude the help of our students Kai-Yuen Cheong, Wai-Man Tam, Ming Ye and Michael Yip in performing the computer simulations. In the course of our research in this area, we have benefited immensely from the stimulating discussions with Prof. Guanrong Chen and Prof. Géza Kolumbán. We also wish to thank our colleagues at the Hong Kong Polytechnic University, especially Simon Hau and Michael Small, for the many hours of fruitful discussions on issues related to chaos and communications. The staff of Springer-Verlag deserve special thanks for their professional and enthusiastic support of this project. Last, but by no means least, we are grateful to the Research Grant Council of Hong Kong and the Research Committee of the Hong Kong Polytechnic University for funding our research work.

Hong Kong
December 2002

Francis C.M. Lau
Chi K. Tse

Table of Contents

1. Introduction

1.1 Overview of Communication Systems

Communication systems, in their most primitive form, are systems that enable or improve the transmission and dissemination of useful information among people in different locations. Technically speaking, the purpose of a communication system is to transmit messages from an information source to one or more destinations. In general, we can identify four basic components in a communication system, as represented by the functional block diagram shown in Fig. 1.1, namely, information source, transmitter, channel, and receiver [Proakis and Salehi (1994); Wozencraft and Jacobs (1965)].

- **Information source:** An information source refers to the origin of information or useful messages which need to be transmitted or disseminated. In modern communication systems, information or messages can take either analog or digital forms. For instance, speech and music are typical analog signals whereas computer data are digital signals. Moreover, analog information can be converted into digital form for the purpose of transmission. In this case, the information is considered digital from the viewpoint of the communication system.
- **Transmitter:** The role of the transmitter is to convert a given message into an appropriate form that is suitable for transmission through the communication channel. In conventional analog transmitters, the message signal is used to modulate (change) the amplitude, frequency or phase of a sinusoidal carrier. Thus, modulation allows the message signal to occupy an allocated frequency spectrum. For digital systems, the modulation of the amplitude, frequency or phase of the sinusoidal carrier by digital symbols constitutes the basic *amplitude-shift-keying* (ASK), *frequency-shift-keying* (FSK) and *phase-shift-keying* (PSK) modulation schemes. In particular, to make efficient use of the bandwidth, the modulation process is designed in such a way that the modulated signal occupies a bandwidth comparable to the data rate. Such communication systems are termed *narrowband systems.*
- **Channel:** The channel is the physical medium through which the signal from the transmitter is transmitted to the receiver. Two basic types of channels can be identified, namely, *wireless* and *wired* channels. Wireless

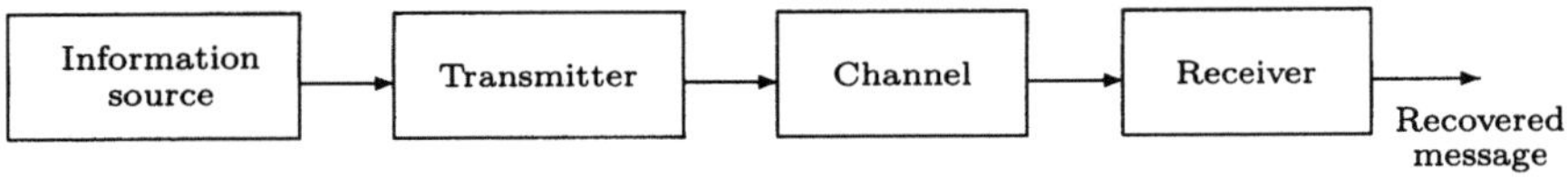

Fig. 1.1. Block diagram of a simplified communication system.

channels are normally the atmosphere, whereas wired channels can take various forms such as twisted pairs, coaxial cables and optical fibres. Regardless of the type of physical medium for signal transmission, the transmitted signal will be corrupted in a random way by a number of possible mechanisms. For example, *additive noise* can be introduced from thermal processes, automobile ignitions, lightnings, etc. Moreover, if more than one user is occupying the same channel at the same time, mutual interference will degrade the signal quality. Another cause of signal degradation due to the channel is multipath propagation, which is the result of signals passing through a number of direct and reflected physical paths between the transmitter and the receiver. Such *distortion* is non-additive and causes time variations in the signal amplitude. The effect is normally called *multipath fading*.

- **Receiver:** At the destination, the original useful message has to be recovered. This is the job of the receiver. Since the signal received has been corrupted by various mechanisms, the message recovery cannot be perfect. The quality of reception is usually described by the *signal-to-noise ratio* (SNR) in the case of the source being in analog form. Moreover, for digital transmission, the quality of reception is measured in terms of the *bit-energy-to-noise-power-spectral-density ratio* (E_b/N_0). The bit energy, normally denoted by E_b, is defined as the average energy transmitted per bit, and $N_0/2$ represents the two-sided power spectral density of an additive white Gaussian noise source. The performance of a digital communication system is usually evaluated in terms of *bit error rate* (BER), which is defined as the probability that a bit will be erroneously decoded. Given two different digital modulation schemes operating at the same data rate, a typical performance comparison is based on their BERs under the same E_b/N_0. Other performance evaluation criteria include *bandwidth efficiency* and *vulnerability to jamming*.

1.2 Spread-Spectrum Communications

Spread-spectrum communications was first developed to provide a secure means of communications for military use under adverse conditions. In a military environment, communications must be protected against two types of attacks, namely, eavesdropping and jamming. Spread-spectrum technology can provide such protections [Peterson *et al.* (1995)].

Unlike a narrowband signal whose power is concentrated within a bandwidth which is roughly equal to the data rate, a spread-spectrum signal "spreads" its power over a much wider bandwidth. As a consequence, the average power spectral density becomes much lower and the signal is easily hidden in the background noise. Without prior knowledge of the communication system or the use of sophisticated equipment, it is not easy to detect the presence of the signal. Even if the presence of the signal is detected, without the appropriate decoding information, the message is hard to recover.

Another problem associated with narrowband systems is that the transmitted signals can be easily jammed or disrupted, for example, by purposely inserting a sinusoidal signal within the bandwidth of the intended narrowband signal. By employing spread-spectrum techniques, the effect of jamming can be substantially reduced, permitting satisfactory recovery of the message. Today, many commercial applications, e.g., mobile communications and positioning systems, also make use of spread-spectrum technologies.

The basic requirements of a spread-spectrum communication system are summarized as follows, according to Peterson *et al.* [1995].

1. The transmitted signal occupies a bandwidth which is much larger than the data rate.
2. The bandwidth of the transmitted signal is independent of the data rate.
3. Demodulation must be achievable, in part, by *correlating* the received signal with a *replica* of the signal used in the transmitter to spread the data.

So far, spread-spectrum techniques have evolved around two main schemes: *direct-sequence* (DS) and *frequency-hopping* (FH). Moreover, there are other spectrum spreading techniques in use, e.g., the time-hopping scheme and some hybrid schemes that combine direct-sequence and hopping methods [Dixon (1994); Peterson *et al.* (1995); Viterbi (1995)]. In the following we give a brief review of the operations of the direct-sequence and frequency-hopping spread-spectrum schemes.

1.2.1 Direct-Sequence Spread-Spectrum Technique

In a direct-sequence spread-spectrum (DS-SS) communication system, the binary data sequence, typically of values ± 1, is first multiplied by a sequence of binary chips, known as *spreading sequence,* which is also of values ± 1. Typical spreading sequences are *m-sequences* and *Gold sequences* [Peterson *et al.* (1995)]. The DS-SS system is designed such that the chip rate is much faster than the data rate. Consequently, the bandwidth of the resulting "spread signal" is determined by the chip rate, which is much larger than the binary data bandwidth. At the receiver, the same sequence of binary chips (now called *despreading sequence*) is reproduced. The despreading process is accomplished by multiplying the despreading sequence with the incoming signal. Based on the correlator output, the binary bit ($+1$ or -1) can be distinguished. If the channel does not distort the signal too badly, and the interference and noise levels are relatively low, the exact data sequence can be recovered.

Figure 1.2 shows the baseband signals in a DS-SS communication system under a distortion-free channel condition. Note that ideal timing synchronization has been assumed and the carrier is omitted for simplicity. The power spectra of the signal before and after spreading are depicted in Fig. 1.3, where the symbol period and chip period are denoted by T_b and T_c, respectively. The ratio of symbol period to chip period, i.e., T_b/T_c, is called the *spreading factor.* In the frequency spectrum, the spreading factor is simply the factor by which the original bandwidth has been expanded. Figure 1.4 shows the block diagram of a particular application of the afore-described DS-SS system which is commonly known as the binary phase-shift-keying (BPSK) DS-SS system.

1.2.2 Frequency-Hopping Spread-Spectrum Technique

Shown in Fig. 1.5 is the block diagram of a frequency-hopping spread-spectrum (FH-SS) communication system. In this system, the signal carrier frequency "hops" among a set of frequencies, according to a pre-determined hopping sequence. In practice, a code generator controls a frequency synthesizer which produces a frequency-hopping signal. Referring to Fig. 1.5, assume that the frequency synthesizer has a hopping frequency of $1/T_c$, i.e., it changes its frequency every T_c seconds, and that the sequence of frequencies is $\{f_1, f_2, f_3, \ldots\}$. The main operation of the FH-SS system is to mix the original binary phase-shift-keying (BPSK) signal with the synthesized carriers, and the resulting signal is let pass through a highpass filter. As a consequence, the BPSK signal at the output of the filter hops according to the carrier sequence $\{f_1 + f_0, f_2 + f_0, f_3 + f_0, \ldots\}$.

At the receiver, a local frequency synthesizer hops in a synchronized manner with the transmitter's hopping sequence to down-convert the received signal to the carrier frequency f_0. The down-converted signal is then demodulated by a conventional demodulator. Thus, without prior knowledge of the hopping sequence, it will be very difficult to decode the information. Moreover, as the signal is now transmitted through a number of carriers, the channel degradation is averaged over all carriers.

For $T_c \geq T_b$ (T_b being the bit period), the system is called a *slow-frequency-hop* system. Otherwise, it is a *fast-frequency-hop* system. In contrast to the case of DS-SS systems, FH-SS signals need not be transmitted in a contiguous frequency band because the carriers $\{f_1, f_2, f_3, \ldots\}$ are not necessarily close to one another. In other words, the resulting frequency spectrum can spread over several disjoint frequency bands.

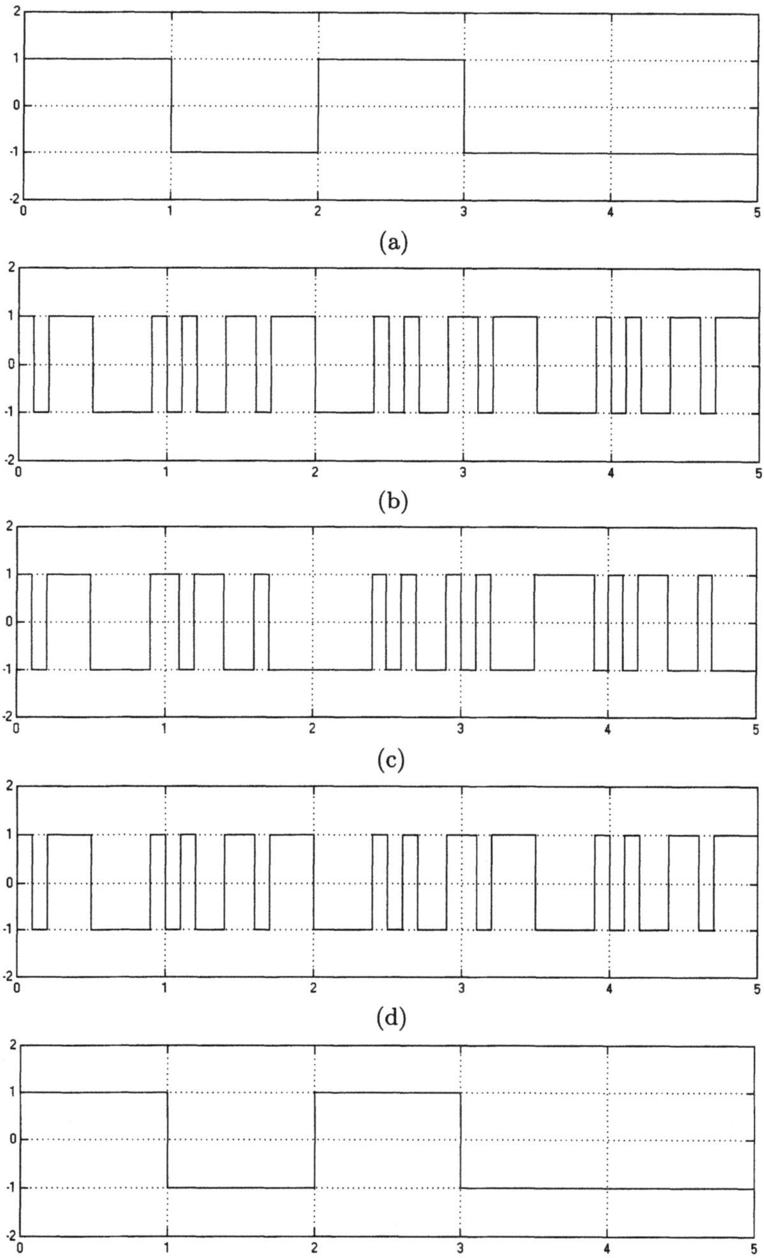

(e)

Fig. 1.2. Baseband waveforms of a direct-sequence spread-spectrum communication system normalized with respect to bit duration. (a) Binary data; (b) spreading sequence; (c) transmitted waveform after spreading; (d) despreading sequence; (e) recovered data.

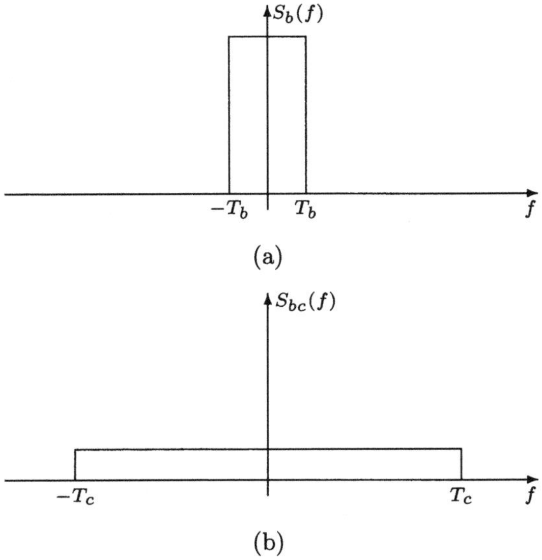

Fig. 1.3. Power spectra of direct-sequence spread-spectrum signals (a) before spreading; and (b) after spreading.

1.3 Advantages of Spread-Spectrum Systems

Obviously, spread-spectrum techniques provide a basic level of security and a good anti-jamming capability. In addition, it can be shown that spread-spectrum communication systems possess several advantages that are not possibly offered by conventional systems. For example, they can mitigate the multipath effect in wireless channels, and in a multiple-access environment, they can average the signal quality of all users. In particular, for mobile cellular networks, spread-spectrum systems reduce the frequency planning effort and increase the system capacity.

1.3.1 Mitigation of Multipath Effects

In a wireless communication environment, usually there exists more than one path (multipath) between the transmitter and the receiver. *Multipath signals* will interfere with one another, causing fades of signals and thus degrading the quality of the receiving signal. For a given bandwidth W of the transmitted signal, it can be shown that multipath components with differential delays of $1/W$ or longer are resolvable [Proakis and Salehi (1994)]. In the DS-SS system, for instance, the enlarged signal bandwidth allows a finer resolution of the multipath components. Once the components are resolved, we can apply RAKE receivers to collect the signal energy from the individual components. This process substantially reduces the effect of multipath and enhances the overall signal quality [Proakis (1995)].

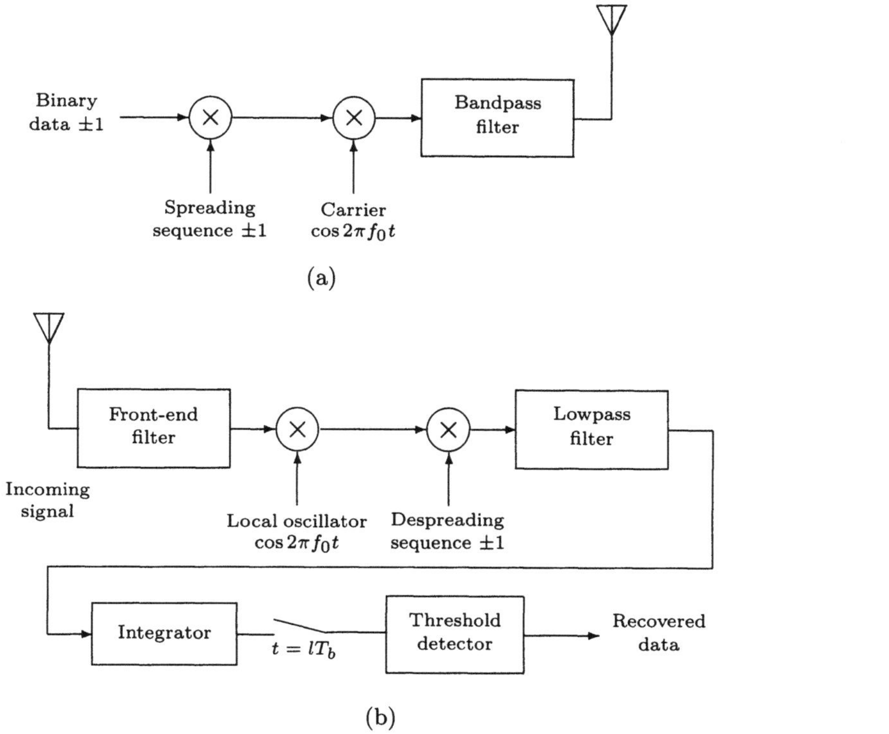

Fig. 1.4. Block diagram of a binary phase-shift-keying direct-sequence spread-spectrum communication system. (a) Transmitter; (b) receiver.

1.3.2 Averaging of Signal Quality in Multiple-User Environments

In the conventional *time-division-multiple-access* (TDMA) or *frequency-division-multiple-access* (FDMA) scheme, each user wanting to transmit a message is assigned a particular time slot or frequency channel, which occupies a narrow bandwidth. Different resources (time slots or frequency channels) may suffer from different levels of distortions depending upon the channel conditions. For example, some resources may suffer from higher interference and may give very poor signal quality.

In DS-SS or FH-SS systems, different users will be assigned with different spreading or hopping sequences. Because all users are using the same bandwidth (on average for FH users) for data transmission, all users will be receiving signals with the same average signal quality.

1.3.3 Reduction of Frequency Planning Effort

In mobile cellular systems, each geographical area is divided into smaller areas called *cells*. To utilize the allocated bandwidth efficiently, different cells

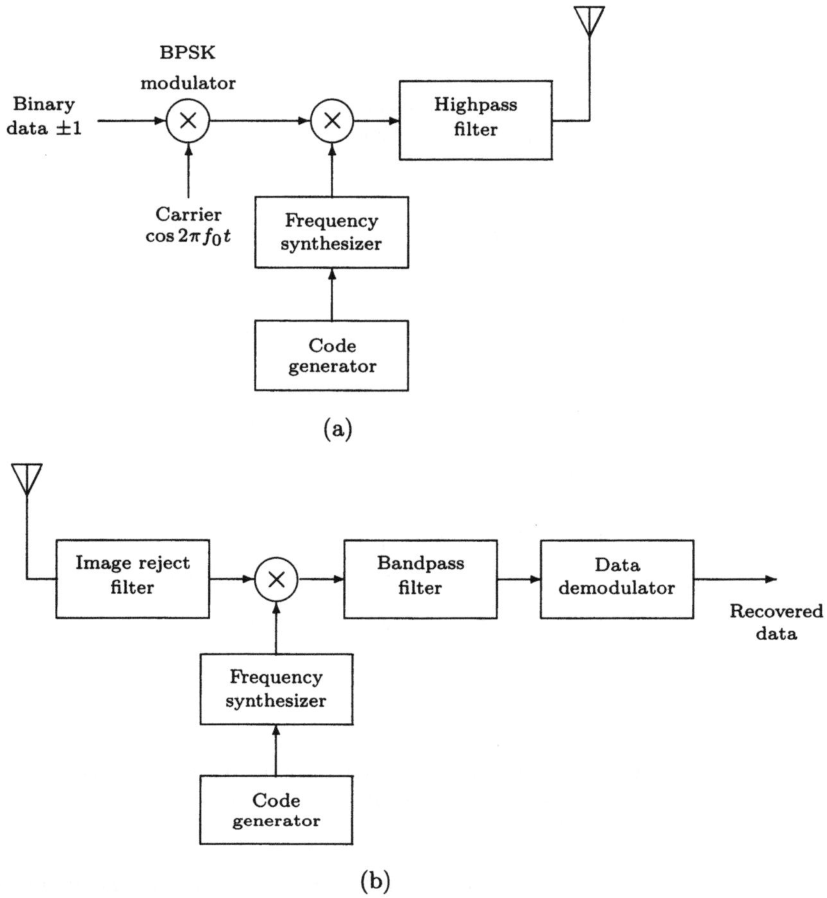

Fig. 1.5. Block diagram of a frequency-hopping spread-spectrum communication system. (a) Transmitter; (b) receiver.

may use the same set of frequency bands at the same time. This is known as *frequency reuse.* In conventional TDMA or FDMA systems, the allocation of frequency bands (carriers) to cells, a process known as *frequency planning,* is usually carried out by experienced engineers who have to ensure that the maximum number of channels are assigned to each cell and that the interference between cells using the same frequency band should be kept below a certain threshold level. Usually, several iterations involving frequency planning and field tests are required, making the frequency planning a rather expensive and time-consuming process.

In a multi-user DS-SS system, each user is assigned with a distinct spreading sequence. This type of multiple access is known as direct-sequence code-division-multiple-access (DS-CDMA). Here, all users are using the same spectrum to transmit their signals at the same time. They are only differentiated by the spreading sequences they use. Therefore, all frequencies can be reused and no cell planning is required [Gilhousen *et al.* (1991)].

1.3.4 Increase in System Capacity

In TDMA or FDMA systems, once all the resources are allocated, no additional users can be accommodated even though some allocated resources may be idle (unused at a certain time). In contrast to TDMA and FDMA systems, DS-SS and FH-SS schemes can flexibly introduce new spreading or hopping sequences whereby more users can be accommodated into the system. The number of spreading or hopping sequences is usually very large. Of course, increasing the number of users will increase the interference level in the case of DS-SS systems and the probability of carrier collision in the case of FH-SS systems. By reducing the transmission power of the users who have no data to send (say, during the silence period in a conversation), the interference caused to others can be reduced. Thus, the overall system capacity can be increased without causing significant signal quality degradation for the users. Analysis shows that DS-CDMA systems can achieve three to seven times the capacity of digital TDMA systems in mobile cellular environments [Gilhousen *et al.* (1991)].

1.4 Applications of Spread-Spectrum Communications

Spread-spectrum techniques are no longer developed solely for military use. Today, the same techniques are applied to a wide range of communication systems for civilian and commercial use. For example, in some second-generation and third-generation mobile communication systems [Garg (1999); Karim *et al.* (2002)], direct-sequence spread-spectrum (DS-SS) is applied to spread the binary symbols in both uplink and downlink. Apart from terrestrial communications, some satellite communication systems also make use of DS-SS as the main modulation method [Elbert (1999)]. Another application that utilizes DS-SS is the Global Positioning System (GPS), also known as NAVSTAR [Clarke (1996)]. In the NAVSTAR system, there are 24 satellites in 12-hour orbits spaced uniformly around the earth and their positions are precisely known all the time. Each satellite sends two DS-SS signals. When a terrestrial user decodes the signals from at least four different satellites simultaneously, the user will be able to calculate his position accurately. Originally developed for military applications, GPS is now widely used in many civilian applications such as geodesic survey, navigation, fleet management, etc.

The most commonly used systems that make use of frequency-hopping spread-spectrum technique include wireless local area network [Santamaria and Lopez-Hernandez (2001)] and the short-range wireless communication protocol Bluetooth [Bray *et al.* (2001)]. Both applications operate in the unlicensed 2.4 GHz band and they support data rates of more than 10 Mbps.

1.5 Chaos-Based Communications

1.5.1 Chaos

Chaotic signals are non-periodic, random-like signals derived from nonlinear dynamical systems [Alligood *et al.* (1996); Chen and Dong (1998); Chen and Ueta (2002); Devaney (1992); Hilborn (2000); Mullin (1993); Strogatz (2001); Thompson and Stewart (2002)]. In general, a dynamical system has a fixed number of independent *state variables,* whose motions or trajectories are governed by a set of differential equations involving all the state variables. For an Nth order system, N state variables exist and a set of N differential equations can be written. To understand how chaotic signals are generated, it suffices to consider discrete-time representations of dynamical systems. Essentially, when a system is described in discrete time, its state variables are sampled at fixed time intervals and its dynamics is described by an *iterative function* which expresses the state variables at one sampling instant in terms of those at the previous sampling instant, i.e., $\boldsymbol{x}_n = \boldsymbol{f}(\boldsymbol{x}_{n-1}, \mu)$, where $\boldsymbol{x}_n$ is the vector of state variables sampled at the nth sampling instant, $\boldsymbol{f}(.)$ is the iterative function that describes the dynamics of the system, and μ is the vector of parameters that affect the system's dynamics.[1]

Chaotic systems are dynamical systems whose state variables move in a bounded, non-periodic, random-like fashion. They are also characterized by a special property known as *sensitive dependence upon initial conditions,* which essentially means that any two nearby starting conditions can lead rapidly to two entirely uncorrelated motions or trajectories of the state variables. This property allows us to generate, theoretically, an infinite number of uncorrelated chaotic signals from the same system using different initial values.

Figure 1.6 (a) shows the waveform of the state variable of a simple first-order discrete-time chaotic system. Here, apart from observing the aperiodic and random nature of the signal, we further confirm that the signal is bounded within the range $[-1, +1]$. To show the sensitive dependence on initial conditions, we plot the waveform again with a slightly different initial value, as shown in Fig. 1.6 (b). Furthermore, it is worth noting that because of their random property, chaotic signals have impulse-like auto-correlation functions and white wideband power spectra. Also, cross-correlation of chaotic signals has a very small value. Typical power spectrum and correlation functions are shown in Figs. 1.7 and 1.8.

[1] The discrete-time representation allows a dynamical system to be studied in terms of an iterative function which takes the current state variables as the input and generates new state variables at discrete times. Thus, the state variables are updated at intervals determined by the sampling period. When the sampling period tends towards zero in the limit, the iterative function resembles a continuous-time representation which takes the form of differential equations.

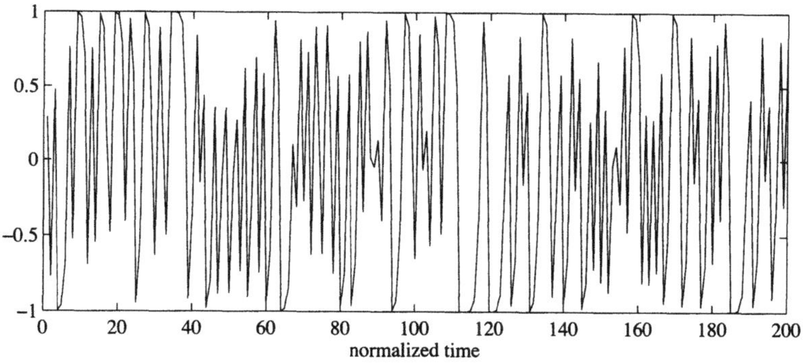

(a)

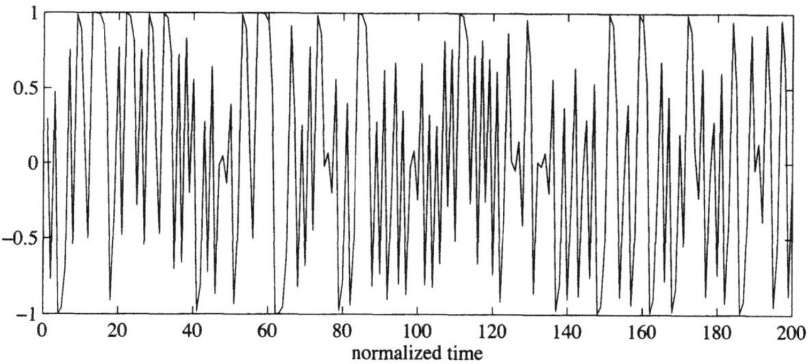

(b)

Fig. 1.6. Waveforms of chaotic signals from a first-order discrete-time dynamical system plotted against normalized time. The generating system is $x_n = 4x_{n-1}^3 - 3x_{n-1}$, which is a cubic map known to be chaotic. Initial value of the system is (a) 0.29; and (b) 0.28999.

1.5.2 Application of Chaos to Communications

Chaotic signals, with their inherent wideband characteristic, are natural candidates for spreading narrowband information. Thus, using chaotic signals to encode information, the resulting signals are spread-spectrum signals having larger bandwidths and lower power spectral densities. They enjoy all the benefits of spread-spectrum signals such as difficulty of uninformed detection, mitigation of multipath fadings, anti-jamming, etc., as mentioned in the previous section. Moreover, a large number of spreading waveforms can be produced easily as a consequence of the sensitive dependence upon initial conditions and parameter variations. Thus, chaos provides a low-cost and versatile means for spread-spectrum communications. In recent years, a number of modulation and demodulation schemes have been proposed for communications. In the following, we briefly review three types of communication techniques that have been widely studied.

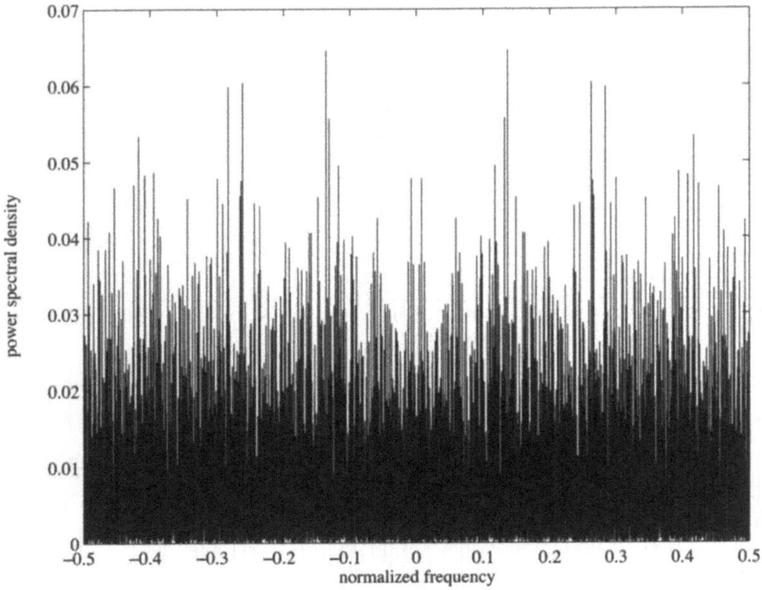

Fig. 1.7. Power spectrum of a chaotic signal against normalized frequency. The generating system is $x_n = 4x_{n-1}^3 - 3x_{n-1}$.

1.5.2.1 Analog Modulation. There are two widely discussed classes of techniques for transmitting analog information with chaotic signals: *chaotic masking* [Cuomo and Oppenheim (1993); Kocarev *et al.* (1992)] and *chaotic modulation* [Itoh and Murakami (1995)]. In the most basic form of chaotic masking, the analog signal is added to the output of a chaotic system. At the receiving end, based on a process called *chaos synchronization,* the original chaotic signal is reconstructed and the analog information is then extracted by subtracting the reproduced chaotic signal from the incoming signal. Moreover, other sophisticated forms of masking have also been proposed, for example, the feedback-based masking [Milanović and Zaghloul (1996)] and the observer-based masking [Liao and Huang (1999)].

The basic idea of chaotic modulation is to inject analog information into a chaotic system, altering its dynamics. This is usually accomplished by modulating a suitably chosen *parameter.* As a result, the chaotic signal generated by the system contains the analog information. The job of the receiver is therefore to track the change in dynamics of the chaotic signal and to retrieve the analog information [Feng and Tse (2001)].

1.5.2.2 Digital Modulation. Several schemes have been proposed in the past for encoding digital information with chaotic signals. In most proposed methods, the basic principle is to map digital symbols to nonperiodic chaotic basis signals. For instance, *chaotic switching* or *chaos shift keying* (CSK) maps different symbols to different chaotic basis signals, which are produced from

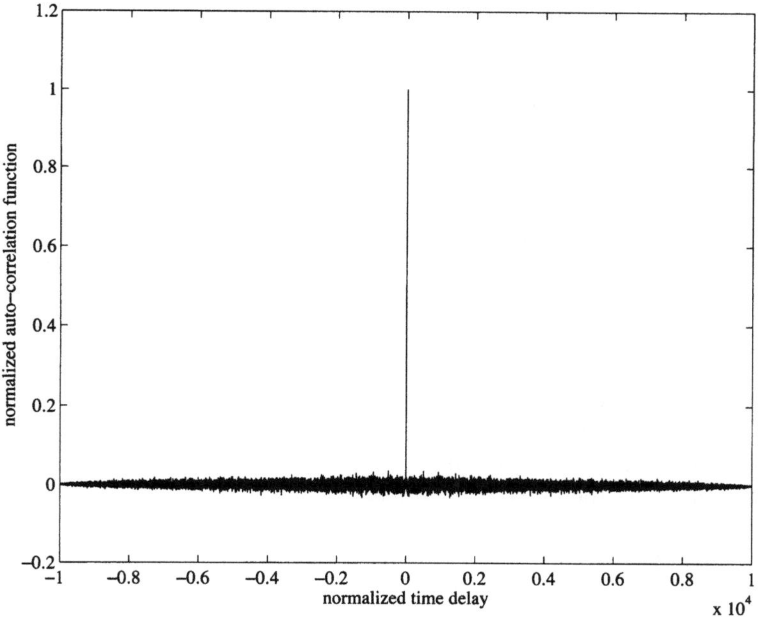

(a)

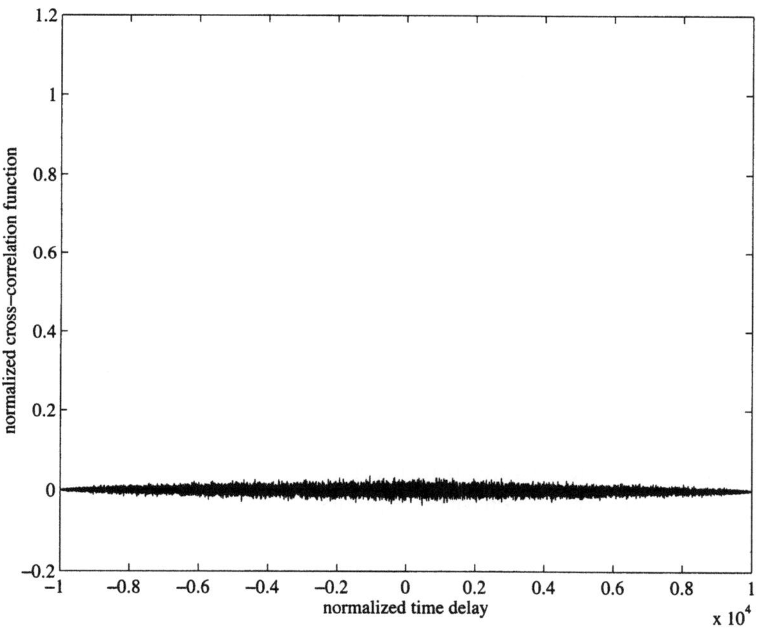

(b)

Fig. 1.8. Normalized correlation functions of chaotic signals against normalized time delay. The generating system is $x_n = 4x_{n-1}^3 - 3x_{n-1}$. (a) Auto-correlation; (b) cross-correlation.

a dynamical system using different values of a bifurcation parameter or from several different dynamical systems. If synchronized copies of the chaotic basis signals are available at the receiver, detection can be achieved by evaluating the synchronization error [Dedieu *et al.* (1993); Parlitz *et al.* (1992)] or based on a conventional correlator-type detector [Kolumbán *et al.* (1998a)]. This class of detection is known as *coherent detection.* [2] Moreover, if synchronized copies of the chaotic basis signals are not available at the receiver, detection has to be done by non-coherent means [Kennedy and Kolumbán (2000)].

Another widely studied modulation technique for encoding digital information is based on a differential keying approach. Known as *differential chaos shift keying* (DCSK) [Kolumbán *et al.* (1996)], this technique essentially fabricates a special structure of the information bit that allows detection to be done in a non-coherent manner, i.e., without the presence of the synchronized copies of the chaotic signals at the receiver. Specifically, in the binary case, every transmitted symbol is represented by two chaotic signal sample sets. The first one serves as the *reference sample set* and the second as the *data sample set.* Dependent upon the symbol being sent, the data sample set is either an exact or inverted copy of the reference sample set. Demodulation can be done in a straightforward manner by correlating the two chaotic sample sets. The binary symbols can be differentiated by comparing the correlator output with a threshold value.

A few other digital modulation schemes derived from CSK and DCSK have also been proposed, e.g., chaotic on-off-keying (COOK) [Kolumbán *et al.* (1997b)], frequency-modulated DCSK (FM-DCSK) [Kolumbán *et al.* (1997d)], correlation delay shift keying (CDSK) and symmetric CSK [Suschik *et al.* (2000)], and quadrature CSK [Galias and Maggio (2001)].

1.5.2.3 Direct-Sequence Spread-Spectrum. Direct application of chaos to conventional direct-sequence spread-spectrum system was first reported by Heidari-Bateni and McGillem [1992]. The basic principle is to replace the conventional binary spreading sequences, such as m-sequences or Gold sequences, by the chaotic sequences generated by a discrete-time nonlinear map. It has been shown that the performance of the new system is comparable to that of the conventional systems using binary spreading sequences [Heidari-Bateni and McGillem (1994)]. The advantages of using chaotic spreading sequences are that an infinite number of spreading sequences exist and that the spread signal is less vulnerable to interception.

Instead of applying analog chaotic sequences to spread the data symbols, Mazzini and his colleagues proposed quantizing and periodically re-

---

[2] Throughout this book, we distinguish coherent detection methods from non-coherent ones by the need for the availability of synchronized replicas of the chaotic signals at the receiver. Thus, when studying coherent detection, we need to assume that synchronized replicas of the chaotic signals are available at the receiver through a chaos synchronization process. However, chaos synchronization is a profound technical subject in physics and mathematics, and is outside the scope of this book.

peating a slice of a chaotic time series for spreading [Mazzini *et al.* (1997); Mazzini *et al.* (1998)]. Analysis shows that systems using the periodic quantized sequences have larger capacities and lower bit error rates than those using m-sequences and Gold sequences in a multiple-access environment.

1.6 Benefits and Challenges

As mentioned in Sect. 1.5.1, chaotic signals are characterized by their random-like nature and broad power spectra. Thus, by embedding information in chaotic signals, it is possible to enhance the data security substantially. Being broadband, chaotic signals, when used for transmission, inherit the advantages of conventional spread-spectrum signals such as low probability of detection, anti-jamming, mitigation of multipath, etc. In addition, chaotic signals are easy to generate, and in theoretically infinite quantity, thus providing a comparatively low-cost solution to the spread-spectrum system construction. In particular, for direct-sequence code-division-multiple-access systems, it has been shown theoretically that system performance in terms of capacity and bit error rates improves when the spreading sequences are generated by quantizing and periodically repeating slices of chaotic time-series.

However, chaos-based communication systems are still considered immature from the practical engineering standpoint, and there are many important technical issues which need to be resolved. In the following we attempt to mention a few challenges in the development of practical chaos-based communication systems.

In practice, the bandwidth of the transmitted signal must always be known. To the best of our knowledge and up to the time of writing, no formal techniques seem to exist for estimating the bandwidths of chaotic signals. Computer simulations become the only resort.

As mentioned earlier, coherent detection requires the availability of synchronized replicas of the chaotic signals at the receiver. This in turn requires robust synchronization between the chaotic systems at the transmitter and the receiver. The problem is non-trivial, and at the time of writing, there are still no acceptable chaos synchronization schemes that can be applied to communication systems for the required low signal-to-noise conditions. This also motivates the research of non-coherent detection which requires no chaos synchronization between the transmitter and the receiver.

Non-coherent detection methods can be regarded as relatively unexplored because of the vast possibilities of detection schemes that can potentially be used. So far, DCSK has been the most widely studied among the non-coherent digital schemes. However, DCSK makes no use at all of the deterministic properties of chaos. It can be envisaged that non-coherent detection will open up new research opportunities in chaos-based communications.

Non-ideal channel conditions, such as distortion and noise, are important concerns in the design of communication systems. When chaotic signals are employed for transmission, the effect of channel distortion can be more profound since it may make synchronization completely unachievable at the receiver [Feng *et al.* (2002)]. Furthermore, the problem of multipath propagation has only been considered for a few selected systems [Kennedy *et al.* (2000)]. More work is certainly needed.

Finally, every chaotic signal is different. The question of how to choose the "most desirable" chaotic signals for use in communication systems remains open. At present, time-consuming simulations are necessarily performed to establish certain relevant properties for analysis such as means, variances, probability density functions, etc.

1.7 What Is This Book About?

The rest of the book is devoted to generic types of chaos-based digital communications. Our emphases are the operating principles, analytical methods and performance evaluation.

Before we embark on a detailed analysis of these systems, we present in the next chapter the salient concepts of chaos-based digital modulations and introduce the operations of some selected types of systems that have been recently proposed.

Beginning in Chap. 3, through Chap. 8, we will take an in-depth investigation of the most widely studied chaos-based digital communication systems, describe in detail the analytical approaches for studying these systems, and present some results of evaluation in a few important performance areas such as anti-jamming and tolerance to mutual interference with existing communication systems.

In the closing chapter we present some ideas for chaos-based communications, with emphasis on non-coherent and non-correlator-type detection. The exposition is surely incomplete, given the vast possibilities in the exploitation of chaos for communications. Nonetheless, our aim is to arouse further interests in research in this emerging discipline.

2. Chaos-Based Digital Modulation and Demodulation Techniques

In this chapter we introduce chaos-based digital communications. Our discussion begins with the conventional digital communication systems. We first review the basic operation of digital modulation, which is to represent digital symbols by periodic *basis signals.*[1] In our main discussion, this same basic operation is extended to chaos-based digital communication systems, in which chaotic basis signals are used in lieu of periodic basis signals in the modulation process. In this chapter we present a summary of the recent development in this field and review a few most widely studied chaos-based digital modulation schemes.

2.1 From Conventional to Chaos-Based Digital Communications

In digital communication systems, the information (digital symbols) to be transmitted is first mapped to some analog basis signals. The process is called *modulation.* The resulting modulated signals are then transmitted through some appropriate channels to the destinations. However, practical channels are non-ideal and have some undesirable effects on the signals being transmitted, e.g., attenuation, bandpass filtering, nonlinear distortion, and noise contamination. All these non-idealities will corrupt the signals and make the job of recovering the original information difficult. Thus, the receiver has to determine the identity of the digital symbols based on the corrupted versions of the basis signals.

Typically, in conventional digital communication systems, the analog basis signals representing the digital symbols differ in either amplitude (as in amplitude shift keying), frequency (as in frequency shift keying) or phase (as in phase shift keying). The receiver detects the differences and decodes the received signals. For the same transmitted symbol, the same analog basis signal is transmitted, except for the case of differential encoding schemes.

In general, we may categorize detection methods into two types, namely, *coherent* and *non-coherent* types. For coherent detection, replicas of the basis

[1] In much of the literature, the term "basis functions" are often used in lieu of "basis signals". Throughout this book, the two terms are used interchangeably.

signals (unmodulated carriers) must be reproduced at the receiver and are used for demodulation. For non-coherent detection, the basis signals need not be recovered at the receiver. For example, in *differential phase shift keying* (DPSK), the signal in one signaling interval is deliberately related in a certain way with that in the preceding signaling interval. Thus, the problem of demodulation is to identify the type of relationship between the signal in one signaling interval and that in the preceding interval [Proakis and Salehi (1994)].

If we use *chaotic basis signals* in lieu of the conventional periodic basis signals for digital communications, the resulting systems are called *chaos-based digital communication systems.* As can be anticipated from the very nature of chaos, the consequence of using chaotic basis signals would lead to profound conceptual differences in the resulting systems. To begin with, the sensitive dependence upon initial condition of chaotic signals will clearly make the reproduction of replica basis signals in the receiver a non-trivial task. As such, coherent detection can be heavily handicapped. However, because coherent detection should theoretically outperform non-coherent detection, the study of coherent systems would provide benchmarks for performance evaluation. For this reason, we study coherent systems to some depth in this book. Moreover, even if the availability of replica basis signals in the receiver is assumed, the study of chaos-based communication systems is still difficult because the usual treatment for periodic signals may not be applicable to chaotic signals, for example, in obtaining statistical means, variances, correlation functions, probability density functions, etc. Thus, the conventional statistical analysis does not necessarily yield analytical expressions for performance evaluation and computer simulations are the only resort. The same problem occurs in the analysis of non-coherent systems as well. In the later chapters of this book, we describe some analytical methods for evaluating the performance of chaos-based digital communication systems. It should be noted that other methods have also been proposed for the study of chaos-based communication systems [Lawrance and Ohama (2002)] and we limit ourselves in this book to analytical methods which the communication engineers are familiar with.

In the following we take a quick tour of the various chaos-based communication systems and attempt to present a way to classify such systems in order to facilitate a systematic investigation into their operations and performances.

2.2 Classifications of Chaos-Based Communication Systems

As in the conventional case, we may classify chaos-based communication systems into two main categories, namely, coherent and non-coherent systems.

Table 2.1. Classification of chaos-based communication systems

Class	System		Correlator-type detection applicable	Some reported systems
Coherent	Analog	Chaotic masking	No	Kocarev *et al.* [1992] Cuomo and Oppenhiem [1993] Milanovic and Zaghloul [1996]
	Digital	*Generic:*		
		Chaos shift keying (CSK)	No	Parlitz *et al.* [1992] Dedieu *et al.* [1993]
		CSK (correlation)	Yes	Kolumbán *et al.* [1998a]
		Symmetric CSK	Yes	Sushchik *et al.* [2000]
		Direct-sequence spread-spectrum:		
		Chaotic spreading sequence	Yes	Heidari-Bateni and McGillem [1992]
		Chaotic digital CDMA	Yes	Yang and Chua [1997]
		Quantized chaotic spreading sequence	Yes	Mazzini *et al.* [1998]
Non-coherent	Analog	Chaotic modulation	No	Itoh-Murakami [1995] Leung and Lam [1997]
		Signal-reconstruction-based system	No	Anishchenko and Pavlov [1998] Feng and Tse [2001]
	Digital	Differential CSK (DCSK)	Yes	Kolumbán *et al.* [1996]
		FM-DCSK	Yes	Kolumbán *et al.* [1997d]
		Chaotic on-off-keying	No	Kolumbán *et al.* [1997b]
		CSK (bit-energy)	No	Kolumbán *et al.* [1998a]
		CSK (optimal)	No	Hasler and Schimming [2000]
		CSK (regression)	No	Tse *et al.* [2001]
		Correlation delay shift keying	Yes	Sushchik *et al.* [2000]
		Quadrature CSK	Yes	Galias and Maggio [2001]
		Ergodic CSK	No	Leung *et al.* [2002]

In coherent systems, the receiver needs to reproduce the chaotic carrier for demodulation, whereas in the non-coherent system, demodulation is done solely based on the received signal.

In Table 2.1, we present a classification of chaos-based communication systems, in which we make distinction between coherent and non-coherent systems, analog and digital systems, and correlator-type and non-correlator-type receivers. The following are worth noting:

- Some schemes can be realized in both coherent and non-coherent forms, e.g., chaos shift keying.
- Although correlator-type receivers are employed in some schemes, their use should represent generic types of detection for coherent systems because correlation operation is a generic process for evaluating the closeness (likeness) between the transmitted signal and the basis signals. In the case of non-coherent systems, however, detection methods can take any form.

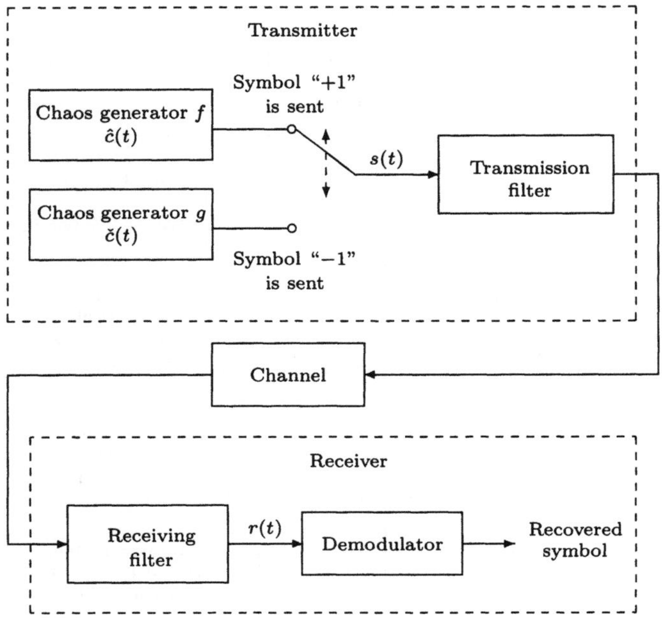

Fig. 2.1. CSK digital communication system.

2.3 Chaos Shift Keying

Chaos shift keying (CSK) was first proposed by Parlitz *et al.* [1992] and Dedieu *et al.* [1993]. As mentioned before, the idea is to encode digital symbols with chaotic basis signals. Figure 2.1 shows the block diagram of a typical CSK digital communication system. The operating principle can be described as follows. The transmitter consists of two chaos generators f and g, producing signals $\hat{c}(t)$ and $\check{c}(t)$, respectively. If a binary "+1" is to be sent during the interval $[(l-1)T_b, lT_b)$, $\hat{c}(t)$ is transmitted, and if "−1" is to be sent, $\check{c}(t)$ is transmitted. In its originally proposed form, the CSK system works on the basis of the self-synchronizing property of chaotic systems. The receiver structure is shown in Fig. 2.2, in which the incoming signal is used to drive two self-synchronization subsystems $\tilde{f}$ and $\tilde{g}$, which are matched to f and g, respectively. Assume that the filters at the transmitter and receiver are distortionless and the channel is perfect. When the transmitted signal is $\hat{c}(t)$, the subsystem $\tilde{f}$ will be synchronized with the incoming signal while $\tilde{g}$ does not, and vice versa. Therefore, by measuring the difference (error) between the incoming signal and the output of the self-synchronization subsystems, the transmitted symbol can be estimated. Results show that the systems work well under a noiseless condition.

In communications, *correlation* is a generic process that is used to evaluate the "likeness" between two signals. Clearly, for the CSK system mentioned above, instead of measuring the synchronization error, we may directly eval-

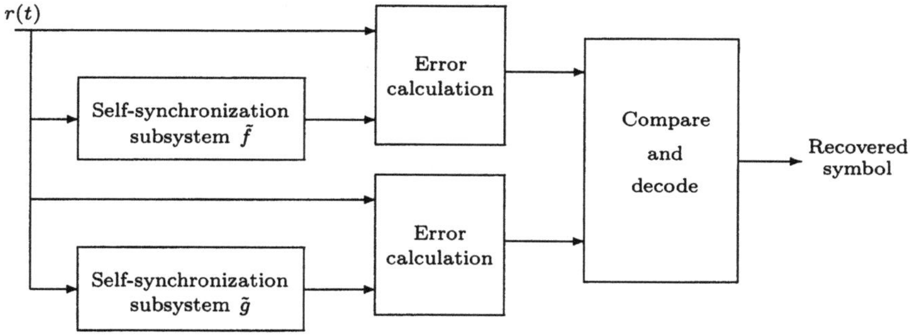

Fig. 2.2. Synchronization-error-based CSK demodulator.

uate the correlation between the transmitted signal and the replica basis signals to identify the transmitted symbol. Thus, a correlator plus a decision maker form a generic coherent receiver for the CSK system.

2.3.1 Coherent Demodulation Based on Correlation

Coherent detection of CSK signals using correlator-based receivers was studied in detail by Kolumbán *et al.* [1998a]. Figure 2.3 shows the block diagram of a correlator-based coherent CSK demodulator. The two synchronization circuits attempt to recover the two chaotic signals $\hat{c}(t)$ and $\check{c}(t)$ from the received corrupted signal $r(t)$. An acquisition time T_s is assumed for the synchronization blocks to lock to the incoming signal. The reproduced chaotic functions are then used to correlate with the received signal during the remainder of the bit duration. Then, the outputs of the correlators are sampled and compared.

Consider the lth received symbol. The outputs of the correlators at the end of the symbol period are given by

$$\hat{o}(lT_b) = \int_{(l-1)T_b+T_s}^{lT_b} r(t)\hat{c}(t)\, dt \tag{2.1}$$

$$\check{o}(lT_b) = \int_{(l-1)T_b+T_s}^{lT_b} r(t)\check{c}(t)\, dt. \tag{2.2}$$

The input to the threshold detector at this instant is

$$y(lT_b) = \hat{o}(lT_b) - \check{o}(lT_b). \tag{2.3}$$

If $y(lT_b)$ is positive, a "+1" is decoded for the lth symbol. Otherwise, a "−1" is decoded. To gain some insights into the demodulation process, the histograms of $y(lT_b)$ are helpful. Assuming that the filters are distortionless and the synchronization time T_s is negligible compared to the bit period, we may plot the histograms of $y(lT_b)$ for different signal-to-noise ratios (SNRs),

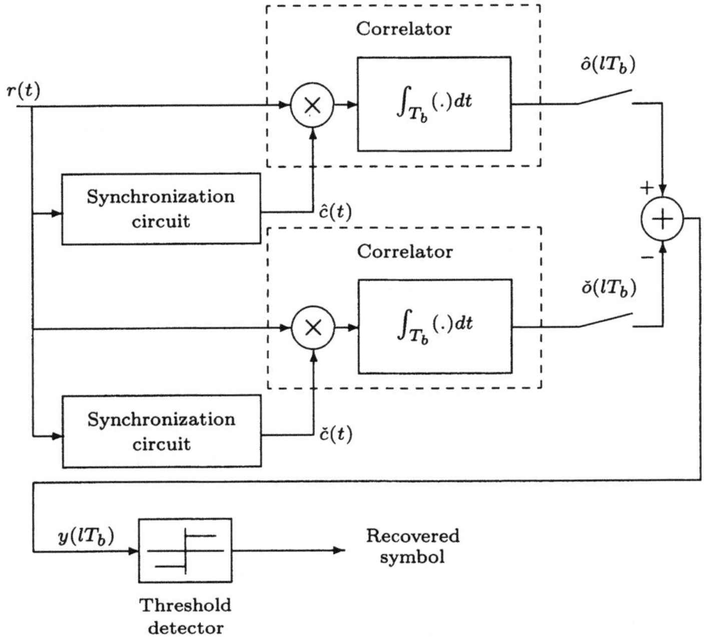

Fig. 2.3. Correlator-based coherent CSK demodulator.

as shown in Fig. 2.4. Clearly, due to the non-periodic nature of the chaotic signal, $y(lT_b)$ varies even for the same symbol. When the SNR is high, the two distinct regions in the histogram guarantee that the transmitted symbols can be decoded correctly. However, when the SNR is low, the two regions overlap and errors become inevitable.

2.3.2 Non-Coherent Demodulation Based on Bit Energy Calculation

In non-coherent CSK demodulation, the chaotic basis signals are not available at the receiver. Detection has to be done based on some distinguishable property of the basis signals. One such property is the bit energy which can be deliberately made different for different symbols in the modulation process [Kolumbán *et al.* (1998a)].

Suppose chaotic basis signals with different bit energies are used to represent the binary symbols. If a binary "+1" is to be sent during the interval $[(l-1)T_b, lT_b)$, a chaotic basis signal $\hat{c}(t)$ with mean bit energy $\hat{E}_c$ is transmitted, and if "−1" is to be sent, a chaotic basis signal $\check{c}(t)$ with mean bit energy $\check{E}_c$ is transmitted. To generate the chaotic signals with different bit energies, we may employ two chaos generators with different average bit energies. Alternatively, the same chaos generator can be used to produce two

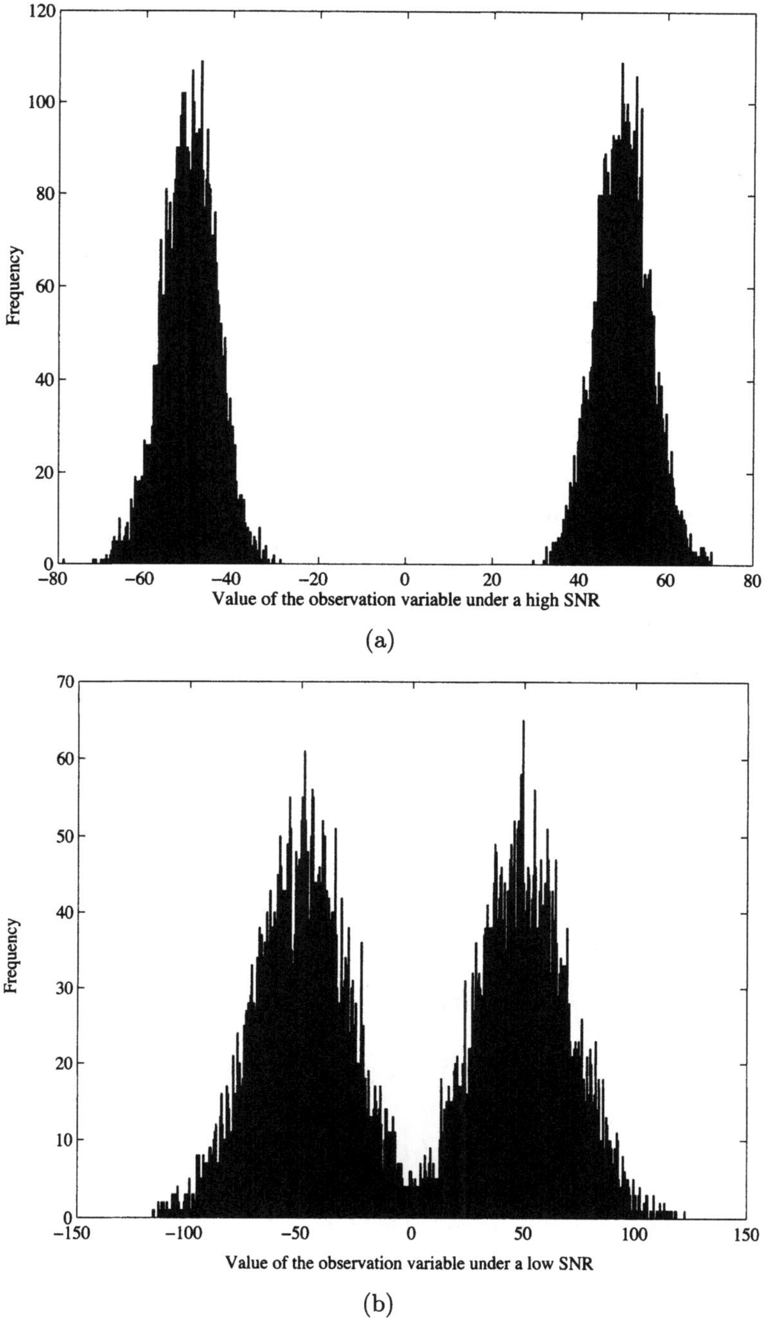

Fig. 2.4. Histograms of the observation variable $y(lT_b)$ for a coherent CSK system for (a) high SNR, and (b) low SNR.

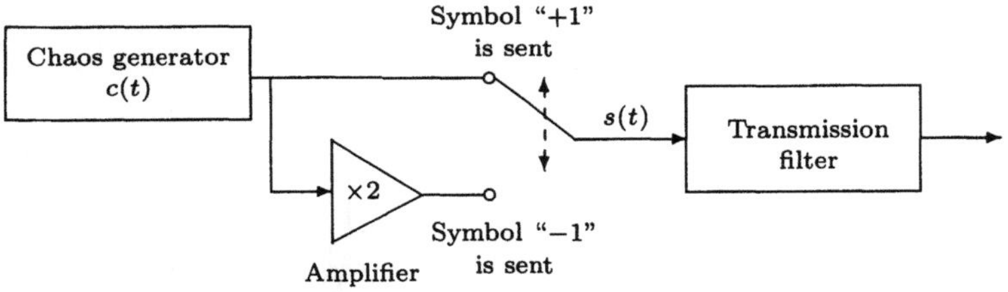

Fig. 2.5. CSK transmitter sending different average bit energies for different symbols.

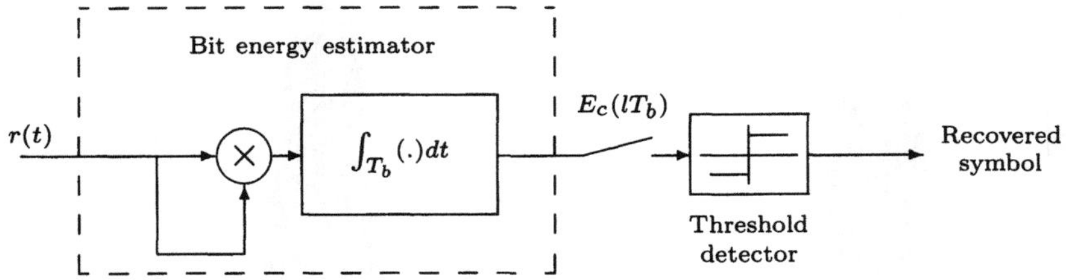

Fig. 2.6. CSK receiver based on bit energy estimator.

signals of different bit energies by using two amplifiers of different gains, as shown in Fig. 2.5. At the receiving end, the bit energy can be estimated by a square-and-integrate process such as the one shown in Fig. 2.6. Assume that only additive noise corrupts the transmitted signal and the noise power is limited by the receiving filter, i.e.,

$$r(t) = s(t) + n'(t) \tag{2.4}$$

where $s(t)$ denotes the transmitted signal and $n'(t)$ is the noise component at the output of the receiving filter. For the lth received symbol, the sampled output of the correlator, or equivalently the received bit energy $E_c(lT_b)$, is given by

$$\begin{aligned} E_c(lT_b) &= \int_{(l-1)T_b}^{lT_b} r^2(t)\, dt \\ &= \int_{(l-1)T_b}^{lT_b} s^2(t)\, dt + 2\int_{(l-1)T_b}^{lT_b} s(t)n'(t)\, dt \\ &\quad + \int_{(l-1)T_b}^{lT_b} [n'(t)]^2\, dt. \end{aligned} \tag{2.5}$$

When the SNR is high, the second and third integrals in (2.5) are negligible compared with the first one. Therefore, $E_c(lT_b)$ is approximately equal to either one of the following two bit energies:

$$\hat{E}_c(lT_b) = \int_{(l-1)T_b}^{lT_b} \hat{c}^2(t)\, dt$$

$$\check{E}_c(lT_b) = \int_{(l-1)T_b}^{lT_b} \check{c}^2(t)\,dt. \tag{2.6}$$

In conventional modulation schemes, the bit energy is fixed for a given symbol. However, in the CSK system, due to the non-periodic nature of chaotic signals, the bit energy for the same symbol is time-varying (i.e., varying from bit to bit), as illustrated by the histogram shown in Fig. 2.7 (a) for the high SNR case. Note that $E_c(lT_b)$ does not show up as two distinct values in the histogram, but rather as clusters around the two mean bit energies $\mathrm{E}[\hat{E}_c(lT_b)]$ and $\mathrm{E}[\check{E}_c(lT_b)]$ with certain variances,[2] which can be reduced by increasing the bit duration T_b. By setting the threshold at the mid-way between $\mathrm{E}[\hat{E}_c(lT_b)]$ and $\mathrm{E}[\check{E}_c(lT_b)]$, the received symbols can be decoded correctly.

For a moderate SNR environment, however, the second and third integrals in (2.5) cannot be ignored. The second integral can be positive or negative while the third is always positive. Hence, the calculated bit energies will generally be larger compared to the high SNR case. Figure 2.7 (b) shows the histogram of $E_c(lT_b)$ under a noisy environment. In this case, the calculated bit energies are larger and the two distinct bit-energy regions in the histogram begin to widen and overlap with each other. Hence, errors are inevitable. Obviously, the optimal threshold (mid-way between $\mathrm{E}[\hat{E}_c(lT_b)]$ and $\mathrm{E}[\check{E}_c(lT_b)]$) is dependent upon the SNR, and this *threshold-shift* problem remains the major drawback of this type of bit-energy-based non-coherent CSK system.

2.4 Differential Chaos Shift Keying

The differential chaos-shift-keying (DCSK) modulation scheme was proposed to facilitate non-coherent detection [Kolumbán *et al.* (1996)], and in particular to combat the problem of threshold shift as we have seen for the case of the non-coherent CSK system discussed in Sect. 2.3.2.

Figure 2.8 shows the block diagram of a DCSK modulator. In this scheme, every transmitted symbol is represented by two consecutive chaotic signal samples. The first one serves as the reference (reference sample) while the second one carries the data (data sample). If a "+1" is to be transmitted, the data sample will be identical to the reference sample, and if a "−1" is to be transmitted, an inverted version of the reference sample will be used as the data sample. Usually, the reference sample is sent in the first half symbol period, and the data sample is sent in the second half symbol period. Thus, for the lth symbol period, we have

$$s(t) = \begin{cases} c(t) & \text{for } (l-1)T_b \le t < (l-1/2)T_b \\ c(t-T_b/2) & \text{for } (l-1/2)T_b \le t < lT_b \end{cases} \tag{2.7}$$

[2] The operator $\mathrm{E}[\cdot]$ denotes the mean.

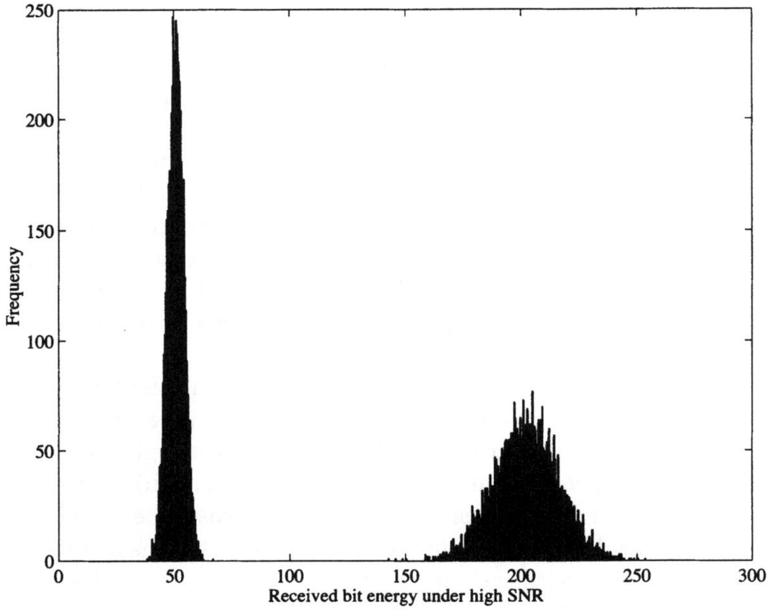

(a)

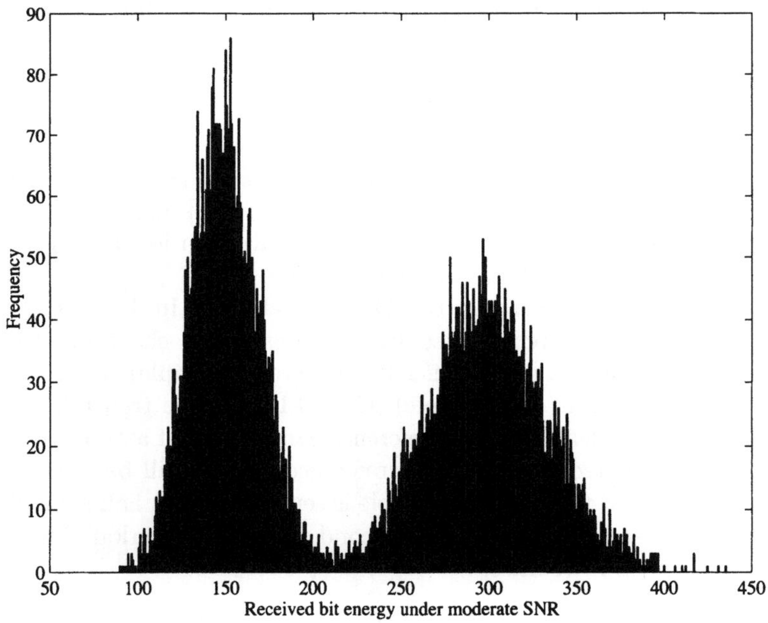

(b)

Fig. 2.7. Histograms of the received bit energy for a non-coherent CSK system for (a) high SNR, and (b) moderate SNR.

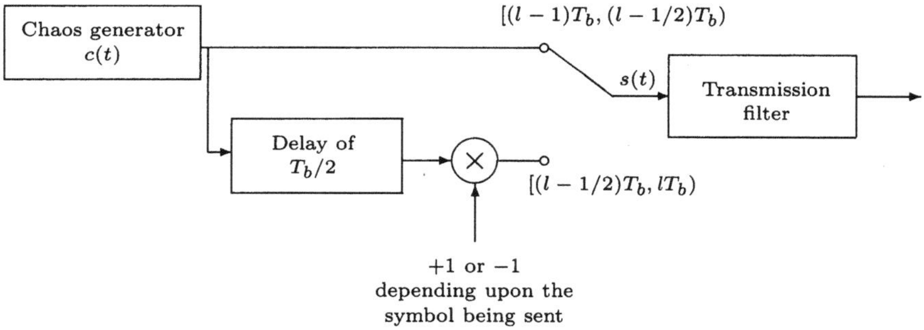

Fig. 2.8. DCSK modulator.

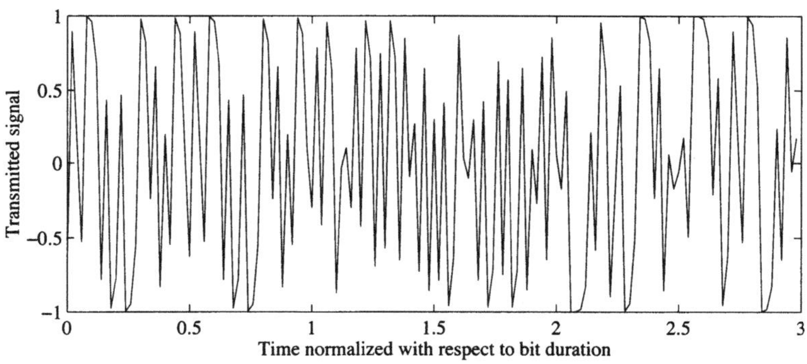

Fig. 2.9. A typical transmitted DCSK signal sample.

if "+1" is to be transmitted, and

$$s(t) = \begin{cases} c(t) & \text{for } (l-1)T_b \leq t < (l-1/2)T_b \\ -c(t-T_b/2) & \text{for } (l-1/2)T_b \leq t < lT_b \end{cases} \tag{2.8}$$

if "−1" is to be sent. Figure 2.9 shows a typical sample of $s(t)$.

At the receiver, the correlation between the reference sample and the data sample is evaluated. This can be done by correlating the incoming signal with a half-symbol-delayed version of itself, as shown in Fig. 2.10. The output of the correlator at the end of the lth symbol duration is given by

$$y(lT_b) = \int_{(l-1/2)T_b}^{lT_b} r(t)r(t-T_b/2)\,dt. \tag{2.9}$$

Assuming that the transmitted signal is contaminated by additive noise, the correlator output is given by

$$y(lT_b) \;=\; \int_{(l-1/2)T_b}^{lT_b} [s(t)+n'(t)][s(t-T_b/2)+n'(t-T_b/2)]\,dt$$

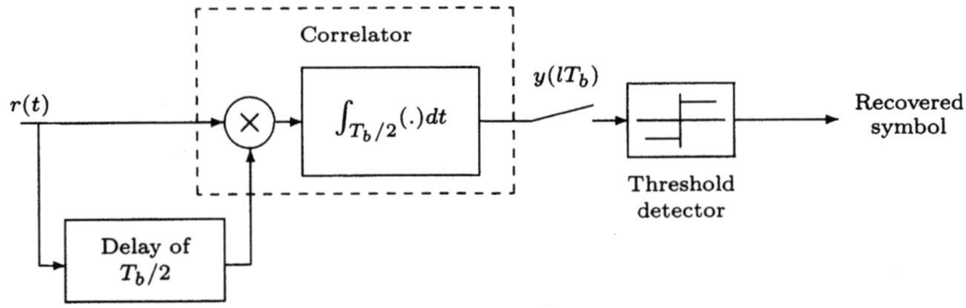

Fig. 2.10. DCSK demodulator.

$$= \int_{(l-1/2)T_b}^{lT_b} [s(t)s(t-T_b/2)]\,dt + \int_{(l-1/2)T_b}^{lT_b} [s(t)n'(t-T_b/2)]\,dt$$
$$+ \int_{(l-1/2)T_b}^{lT_b} [n'(t)s(t-T_b/2)]\,dt$$
$$+ \int_{(l-1/2)T_b}^{lT_b} [n'(t)n'(t-T_b/2)]\,dt \qquad (2.10)$$

where $n'(t)$ is the noise component at the output of the receiving filter. The first term in (2.10) can be positive or negative, depending on whether a "+1" or "−1" has been transmitted. Also, all other integral terms have a zero mean. Thus, the threshold of the detector can be set optimally at zero, which is independent of the noise level.

Figure 2.11 (a) shows the distribution of the correlator output under a high SNR environment. Note that the centres of the two clusters corresponding to the two symbols are located at equal distance from zero. Thus, setting the threshold at zero is sufficient to differentiate the two symbols. If the channel is noisy, however, the two clusters widen and overlap each other, as shown in Fig. 2.11 (b). Thus, although errors are inevitable, the optimal threshold remains at zero. This is clearly an advantage of DCSK over the non-coherent CSK discussed earlier. Furthermore, DCSK is almost insensitive to channel distortion. This is because the channel usually does not vary much within a symbol period and both the reference and data samples are thus subject to the same distortion.

The main drawback of DCSK, however, is that it can only transmit at half the data rate of the other systems because it spends half of the time transmitting the non-information-bearing reference samples. One possible way to increase the data rate is to use a multilevel demodulation scheme [Kolumbán *et al.* (1997c)]. The price to pay, however, is a more complicated system and a possibly degraded bit error performance due to the attenuation of the channel.

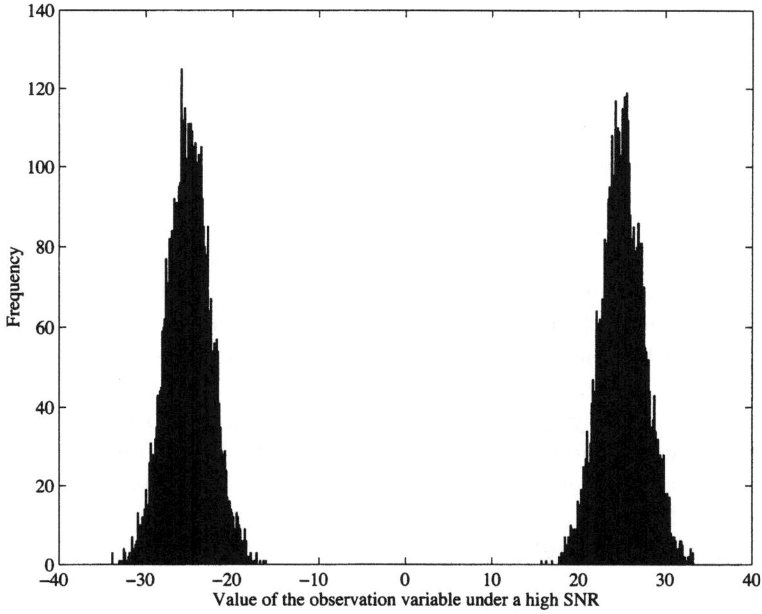

(a)

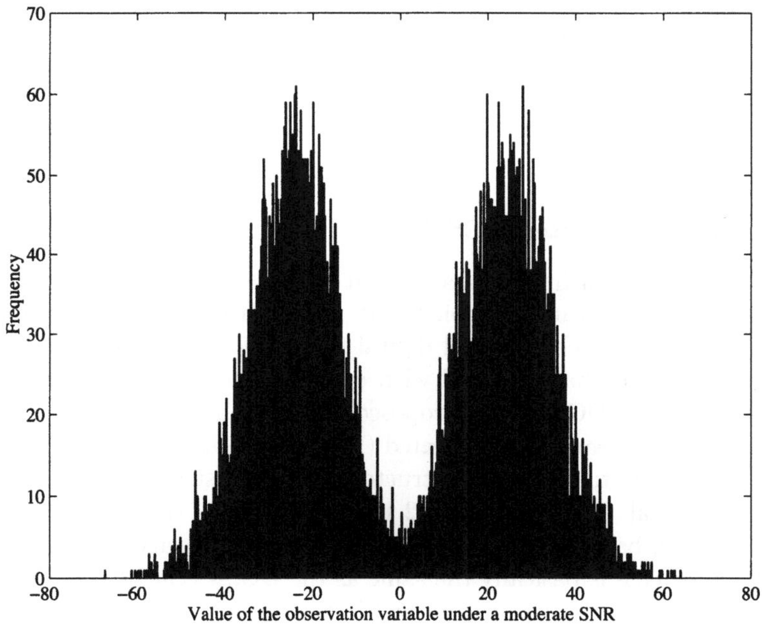

(b)

Fig. 2.11. Histogram of the observation variable $y(lT_b)$ for a DCSK system. (a) High SNR; (b) low SNR.

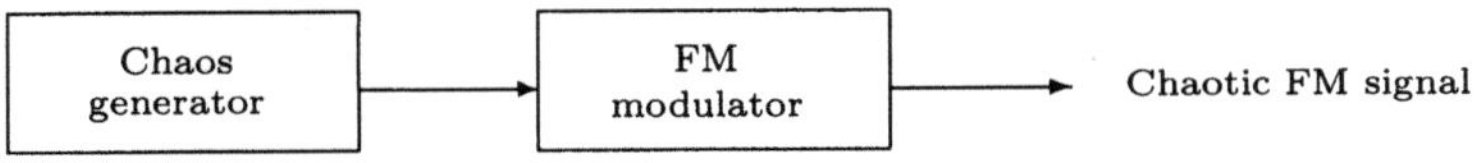

Fig. 2.12. Chaotic frequency-modulated signal generator.

2.5 Other Modulation Schemes

In the foregoing sections, we have reviewed two most widely studied chaos-based digital modulation schemes: CSK and DCSK. The key concept in CSK is to encode digital symbols with chaotic basis signals. Demodulation can be coherent or non-coherent. In DCSK, the structural arrangement of the symbol is exploited to facilitate non-coherent demodulation. Since the introduction of CSK and DCSK, several variations of these schemes were proposed, for example, chaotic on-off-keying, frequency-modulated DCSK, correlation delay shift keying, symmetric CSK and quadrature CSK. In the following we briefly review these CSK- or DCSK-derived modulation schemes.

2.5.1 Chaotic On-Off-Keying

Chaotic on-off-keying (COOK) is a special case of non-coherent CSK. In this scheme, only one chaotic basis signal is needed. The binary symbols "+1" and "−1" are simply represented by transmission of this signal and null transmission, respectively [Kolumbán *et al.* (1997b)]. The modulation process is thus as simple as turning on and off a chaos generator. Clearly, the bit energy of the transmitted signal takes either a positive value or zero, depending on the symbol sent. The demodulation can be done in a non-coherent manner using a simple bit-energy estimator.

2.5.2 Frequency-Modulated DCSK

As illustrated earlier with histograms, the bit energy corresponding to the same binary symbol may vary from bit to bit, and such bit energy variation is an inherent problem in chaos-based digital modulation schemes. In order to produce a wideband chaotic signal with constant power, the frequency-modulated DCSK (FM-DCSK) was proposed [Kolumbán *et al.* (1997d)]. In this scheme, a chaotic frequency modulated (FM) signal generator is needed. This chaotic FM generator can be constructed from the usual FM generator fed by a chaotic signal, as shown in Fig. 2.12. The output of this chaotic FM generator is chaotic, bandlimited, and its power spectral density is uniform. The FM-DCSK scheme is basically the same as the DCSK scheme, with the chaotic basis signals generated from chaotic FM generators. The demodulator is the same as the DCSK demodulator. The histograms of the correlator output from an FM-DCSK demodulator are shown in Fig. 2.13. Note that at high SNR, the histogram shows two distinct outputs corresponding to the two symbols.

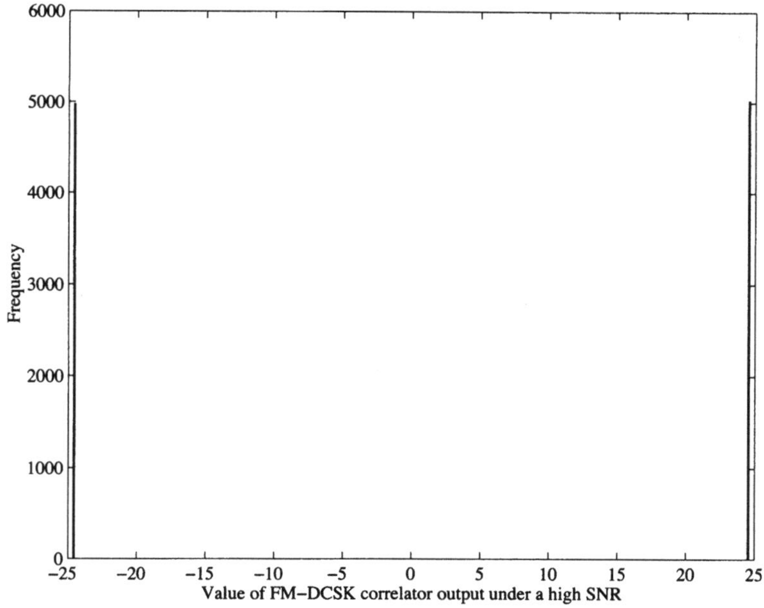

(a)

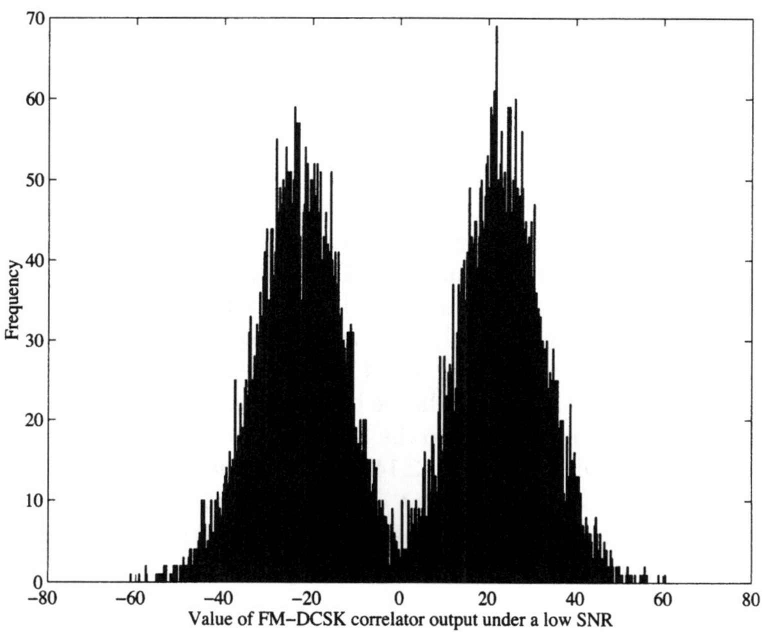

(b)

Fig. 2.13. Histograms of the correlator output for an FM-DCSK system. (a) High SNR; (b) low SNR.

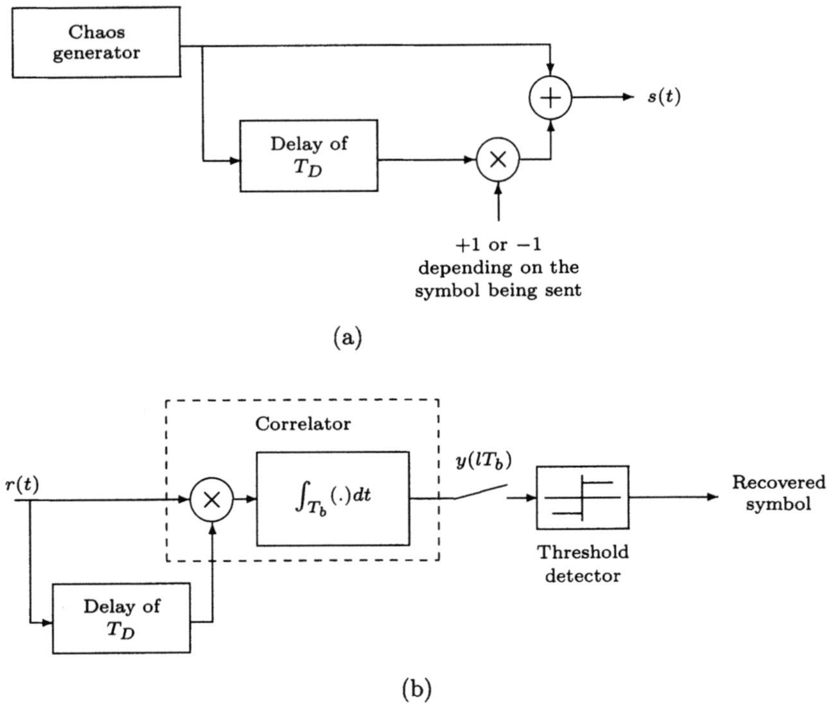

Fig. 2.14. Correlation-delay-shift-keying system. (a) Modulator; (b) demodulator.

2.5.3 Correlation Delay Shift Keying

Correlation delay shift keying (CDSK) can be considered as a derivative of DCSK [Sushchik *et al.* (2000)]. In this scheme, the transmitted signal is the sum of the chaotic signal and its delayed version multiplied by the information signal, as shown in Fig. 2.14 (a). Compared to DCSK, the CDSK scheme does not need to perform any switching and thus allow a continuous operation of the transmitter. Furthermore, the delay T_D is not necessarily equal to half the symbol period and the transmitted signal is never repeated. This leads to a homogeneous transmitted signal and hence better tolerance to interception.

The CDSK demodulator is similar to the DCSK demodulator, except that the delay equals T_D, as shown in Fig. 2.14 (b). Here, we note that due to the non-zero correlation between different chaotic segments, the uncertainty of the correlator output generally increases and the bit error performance degrades in comparison to the DCSK scheme.

2.5.4 Symmetric Chaos Shift Keying

Symmetric chaos shift keying (SCSK) can be considered as a sub-class of antipodal chaos shift keying [Sushchik *et al.* (2000)]. The operation of the

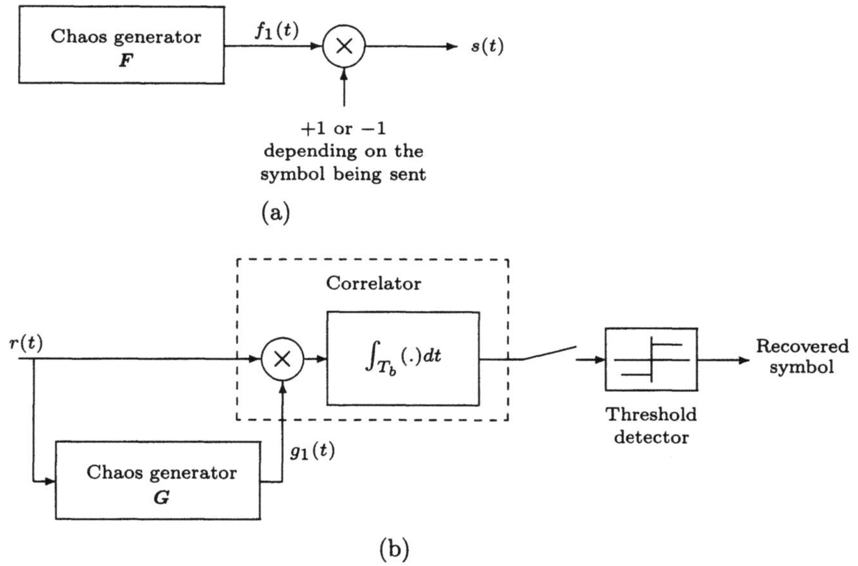

Fig. 2.15. Symmetric CSK system. (a) Modulator; (b) demodulator.

SCSK scheme is depicted in Fig. 2.15. The chaotic system in the transmitter is represented by

$$\dot{\boldsymbol{f}}(t) = \boldsymbol{F}(\boldsymbol{f}(t)) \tag{2.11}$$

where $\boldsymbol{f}(t)$ denotes the internal state vector of the system at time t. The transmitted signal $s(t)$ is obtained from multiplying the first component of $\boldsymbol{f}(t)$, i.e., $f_1(t)$, by the digital symbol being sent.

At the receiver, the signal is used to drive a matched chaotic system

$$\dot{\boldsymbol{g}}(t) = \boldsymbol{G}(|f_1(t)|, \boldsymbol{g}(t)). \tag{2.12}$$

Thus, the systems $\boldsymbol{F}(.)$ and $\boldsymbol{G}(.)$ form a drive-response system with $f_1(t)$ and $g_1(t)$ being the coupling pair. In a distortion-free channel, the output of the chaotic system in the receiver is the same as that in the transmitter, i.e., $f_1(t) = g_1(t)$. The transmitted symbol can thus be obtained by observing the sign of the correlator output at the end of each symbol period.

The SCSK transmitter is simpler compared to DCSK and CDSK. Furthermore, the SCSK signal is non-repeating, resulting a lower probability of interception. Because a matched nonlinear system is required in order to demodulate the SCSK signal, a higher level of security is provided against unauthorized decoding. However, bit error performance is generally worse than that of DCSK and CDSK, and the scheme is restricted by the choice of nonlinear systems.

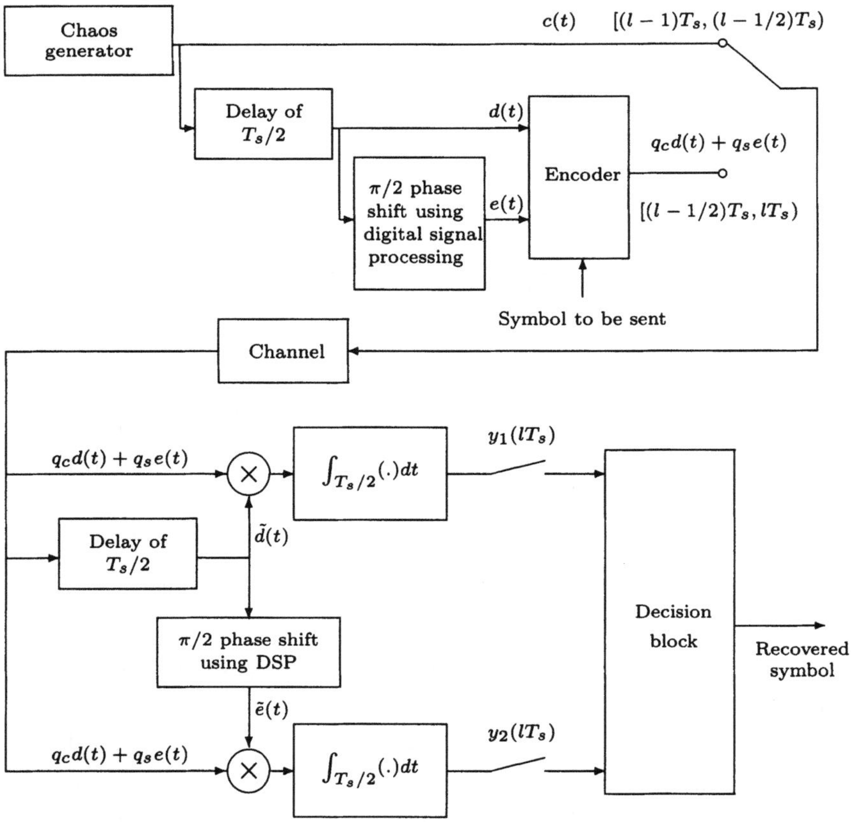

Fig. 2.16. Quadrature CSK system.

2.5.5 Quadrature Chaos Shift Keying

Quadrature chaos shift keying (QCSK) is a multilevel version of DCSK exhibiting a higher spectral efficiency [Galias and Maggio (2001)]. Figure 2.16 shows the block diagram of a QCSK communication system. In this scheme, each transmitted symbol consists of two bits of information. Suppose the bit duration is T_b. Then, the symbol duration T_s is equal to $2T_b$.

The modulation scheme can be described as follows. We first denote by $c(t)$ a chaotic reference signal which is defined in $t \in [0, T_s/2)$. This reference signal has a zero mean. Next, we take the $T_s/2$-delayed version of $c(t)$ to produce $d(t)$. Further, we construct a *complementary signal* $e(t)$ by shifting the phase of all frequency components in $d(t)$ by $\pi/2$, which is accomplished by standard digital signal processing (DSP) techniques. In the QCSK modulation scheme, $c(t)$ is sent during the first half symbol period, i.e., $[0, T_s/2)$. In the second half symbol period, i.e., $[T_s/2, T_s)$, the information-bearing signal $s(t)$ is sent. The construction of $s(t)$ is according to

$$s(t) = q_c d(t) + q_s e(t) \tag{2.13}$$

where q_c and q_s are the two bits of information to be sent within the symbol period T_s.

At the receiving side, $d(t)$ and $e(t)$ are first estimated from the corrupted reference signal $\tilde{c}(t)$. Suppose the estimated $d(t)$ and $e(t)$ are $\tilde{d}(t)$ and $\tilde{e}(t)$, respectively. Then, demodulation can be done by correlating the signal received in the second half symbol period with $\tilde{d}(t)$ and $\tilde{e}(t)$. The symbol (two bits of information) can then be decoded by an appropriate decision making algorithm based on the correlator outputs.

The QCSK can be considered as two DCSK systems. The first one uses the delayed reference signal $d(t)$ and the second one uses the complementary signal $e(t)$ for correlating with the information-bearing signal. The QCSK scheme therefore has doubled the data rate at the expense of a higher system complexity.

2.6 Discrete-Time Baseband Equivalent Models

In the analysis of digital communication systems, the so-called *discrete-time baseband models* are often used. These models permit simple unified analysis to be performed to evaluate the performance of digital communication systems. Throughout this book, we will adopt this type of discrete-time baseband models for the analysis of chaos-based digital communication systems.

In the following, we briefly explain the procedure of transforming a continuous-time bandpass system to its discrete-time baseband equivalent model. Of particular interest is the average *bit-energy-to-noise-power-spectral-density ratio* which is usually defined for the bandpass system. In order to produce consistent analytical results, the bit-energy-to-noise-power-spectral-density ratio has to be re-defined in terms of the parameters of the discrete-time baseband model, as will be explained in the following. Additional information about the modelling process may be found from the work by Proakis and Salehi [1994] and Hasler and Schimming [2000].

2.6.1 Bandpass System

In typical bandpass communication systems, a high-frequency analog signal $s(t)$ is sent from the transmitter, through the channel, to the receiver. Usually, the transmitted signal occupies a bandwidth of W with a centre frequency $f_0 \gg W$. A sketch of a typical spectrum of the transmitted signal is shown in Fig. 2.17. As the signal goes through the channel, its appearance is altered in a number of ways, which are collectively referred to as *distortion*. For example, attenuation, delay spreading, multipath fading, additive noise, etc., distort the signal in different ways. The receiver's job is therefore to recover

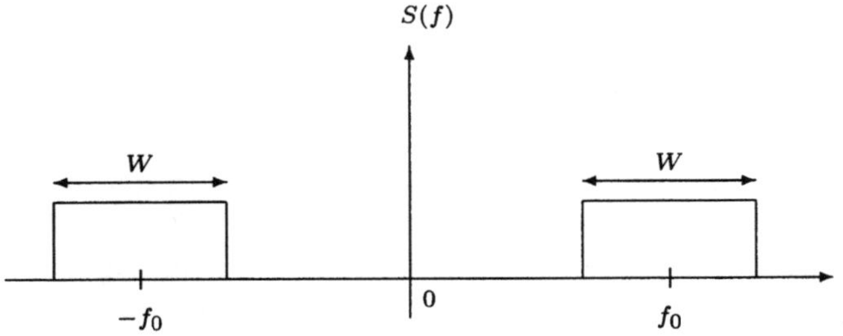

Fig. 2.17. Sketch of spectrum of transmitted signal produced by transmitters in typical bandpass communication systems with centre frequency f_0 and bandwidth W.

the carried information based on the distorted received signal. Here, we restrict ourselves to a simple *additive white Gaussian noise* (AWGN) channel. Precisely, the distortion is caused by an additive noise source $n(t)$ which is assumed to be zero-mean, *wide-sense stationary* Gaussian, with two-sided *power spectral density* $S_{nn}(f)$ and *autocorrelation* $R_{nn}(\tau)$ given, respectively, by,

$$S_{nn}(f) = \frac{N_0'}{2} \tag{2.14}$$

and

$$R_{nn}(\tau) = \mathrm{E}[n(t)n(t+\tau)] = \frac{N_0'}{2}\delta(\tau). \tag{2.15}$$

where $\delta[\cdot]$ is the *impulse function.*

At the receiver, a bandpass filter is first applied to limit the noise bandwidth of the incoming signal. Assuming that an ideal filter of bandwidth W is used, all interfering signals that do not fall within the frequency band occupied by the transmitted signal will be eliminated. The output of the filter is of the form

$$r(t) = s(t) + n'(t) \tag{2.16}$$

where $n'(t)$ denotes the filtered Gaussian noise signal.

The bandpass transmitted signal $s(t)$ is usually produced by modulating a high-frequency carrier with a low-frequency information-bearing signal. Suppose the information-bearing signal is chaotic, e.g., produced by one of those chaos-based modulation schemes discussed earlier. The receiver first retrieves the chaotic information-bearing signal in a conventional manner, then applies the appropriate demodulation scheme to recover the information.

2.6.2 Lowpass Equivalent Model

For most practical purposes, the study of bandpass communication systems is more conveniently carried out in terms of their baseband equivalent systems [Proakis and Salehi (1994)]. The idea is to transform the bandpass system to a lowpass equivalent system on which analysis can be performed more easily and in a more unified way. Basically, if the bandpass signal is denoted by

$$x(t) = x_c(t)\cos(2\pi f_0 t) - x_s(t)\sin(2\pi f_0 t), \tag{2.17}$$

the lowpass representation of $x(t)$, denoted by $x_l(t)$, is given by

$$x_l(t) = x_c(t) + jx_s(t). \tag{2.18}$$

In general, $x_l(t)$ is complex, with real part $x_c(t)$ and imaginary part $x_s(t)$. Both $x_c(t)$ and $x_s(t)$ are lowpass signals, normally called *in-phase* and *quadrature components* of the bandpass signal. Also, the bandpass signal can be recovered from the lowpass signal using

$$x(t) = \Re[x_l(t)\exp(j2\pi f_0 t)] \tag{2.19}$$

where $\Re[\cdot]$ represents the real part operator.

Denoting the lowpass equivalent of $s(t)$, $n'(t)$ and $r(t)$ by $s_l(t)$, $n'_l(t)$ and $r_l(t)$, respectively, we can write

$$r_l(t) = s_l(t) + n'_l(t) \tag{2.20}$$

where

$$r_l(t) \equiv r_c(t) + jr_s(t) \tag{2.21}$$
$$s_l(t) \equiv s_c(t) + js_s(t) \tag{2.22}$$
$$n'_l(t) \equiv n_c(t) + jn_s(t). \tag{2.23}$$

The subscripts c and s denote the in-phase and quadrature components, respectively. The real and the imaginary parts of the lowpass equivalent noise source $n'_l(t)$ are generated from two independent *Gaussian processes*, each having a power spectral density N'_0 and being restricted to the frequency band $[-W/2, W/2]$.

For simplicity, we consider the special case where the lowpass signal $s_l(t)$ is real. Obviously, this implies an incomplete utilization of the frequency band. Under this assumption, we only need to consider the real part of all lowpass signals, i.e., $r_c(t)$, $s_c(t)$ and $n_c(t)$, i.e.,

$$r_l(t) = r_c(t) = s_c(t) + n_c(t). \tag{2.24}$$

2.6.3 Discrete-Time Lowpass Equivalent Model

In digital communications, the information signal is discrete in nature. Typically a bit is sent occupying a fixed duration in time, and its value assumes a few discrete levels (e.g., two levels in the binary case). It is therefore natural to use a discrete approach to study digital communication systems. First, we consider the discrete-time version of the lowpass equivalent model which can be obtained by sampling the equivalent lowpass signal at a frequency W

[Hasler and Schimming (2000)]. Thus, $T_c = 1/W$ is the sampling period. The sampled signal at the receiving filter output is given by

$$r_k \equiv r_c(kT_c) = s_c(kT_c) + n_c(kT_c) \quad \text{for all } k \in I. \tag{2.25}$$

From the Nyquist sampling theorem, the analog lowpass equivalent signal can be recovered from the sampled signal by an ideal lowpass filtering with a bandwidth of $W/2$. For brevity, we denote $s_c(kT_c)$ and $n_c(kT_c)$ by s_k and n_k, respectively. Therefore, (2.25) becomes

$$r_k = s_k + n_k. \tag{2.26}$$

To evaluate the statistical properties of n_k, we need to investigate the lowpass equivalent noise component $n_c(t)$, which is a Gaussian process having a power spectral density N_0' and being restricted to the frequency band $[-W/2, W/2]$. In other words, the noise power spectral density of $n_c(t)$ is given by

$$S_{nc}(f) = \begin{cases} N_0' & |f| \leq \frac{W}{2} \\ 0 & \text{elsewhere.} \end{cases} \tag{2.27}$$

The autocorrelation of $n_c(t)$ can be evaluated by performing an inverse *Fourier transform* on the power-spectral-density function, i.e.,

$$\mathrm{E}[n_c(\tau)n_c(\tau + t)] = F^{-1}[S_{nc}(f)] = WN_0' \operatorname{sinc}(Wt). \tag{2.28}$$

where $F^{-1}[\cdot]$ denotes the inverse Fourier transform operator and $\operatorname{sinc}(x) = \sin(\pi x)/(\pi x)$. Note that the sinc function vanishes for all $t = k/W$ when k is a non-zero integer. Thus, for $m \neq 0$, the correlation of n_k and n_{k+m} is zero because

$$\mathrm{E}[n_k n_{k+m}] = \mathrm{E}\left[n_c(kT_c)n_c((k+m)T_c)\right] = WN_0' \operatorname{sinc}(m) = 0. \tag{2.29}$$

Since n_k and n_{k+m} are uncorrelated Gaussian variables, they are statistically independent [Proakis and Salehi (1994)]. Also, from (2.28), we see that the variance of n_k (noise variance), denoted by σ_n^2, equals WN_0'.

2.6.4 Derivation of Average Bit-Energy-to-Noise-Power-Spectral-Density Ratio

In a digital communication system, the performance is usually evaluated in terms the *bit error rate* (BER) for a particular bit-energy-to-noise-power-spectral-density ratio. For simplicity, we consider a binary system with a bit duration of T_b. To evaluate the average bit energy of the bandpass signal, we make the following assumptions which hold for most practical circumstances:

1. The carrier frequency f_0 is much larger in magnitude than the bandwidth of the signal, i.e., $f_0 \gg W$.

2. The bit duration T_b is an integral multiple of the sampling period T_c. For notational brevity, we let 2β be T_b/T_c.[3]
3. The square of the real lowpass equivalent signal $s_c(t)$, i.e., $s_c^2(t)$, changes slowly within T_c.

With these assumptions and if $s_s(t)$ (imaginary part of $s_l(t)$) equals zero, the average bit energy of the bandpass signal, denoted by E_p, is given by

$$\begin{aligned}
E_p &= \mathrm{E}\left[\int_0^{T_b} s^2(t)\,dt\right] \\
&= \mathrm{E}\left[\int_0^{T_b} s_c^2(t)\cos^2(2\pi f_0 t)\,dt\right] \\
&= \mathrm{E}\left[\int_0^{T_b} s_c^2(t)\,dt\right]\mathrm{E}\left[\cos^2(2\pi f_0 t)\,dt\right] \quad \text{(Assumption 1)} \\
&= \mathrm{E}\left[\int_0^{T_b} s_c^2(t)\,dt\right]/2 \\
&\approx \mathrm{E}\left[\int_0^{T_b} \sum_{k=0}^{2\beta-1} s_c^2(kT_c)p(t-kT_c)\,dt\right]/2 \quad \text{(Assumptions 2 and 3)} \\
&= \mathrm{E}\left[\sum_{k=0}^{2\beta-1} c_k^2\right] T_c/2 \qquad (2.30)
\end{aligned}$$

where $c_k = s_c(kT_c)$ and $p(t)$ is a rectangular pulse of unit amplitude and width T_c, i.e.,

$$p(t) = \begin{cases} 1 & 0 \le t < T_c \\ 0 & \text{elsewhere.} \end{cases} \qquad (2.31)$$

The bit-energy-to-noise-power-spectral-density ratio is thus equal to

$$\zeta \equiv \frac{E_p}{N_0'} = \frac{\mathrm{E}\left[\sum_{k=0}^{2\beta-1} c_k^2\right] T_c}{2N_0'} \qquad (2.32)$$

where N_0' is the power spectral density in the passband.

In the discrete-time lowpass equivalent model, the normalized average bit energy is defined as

$$E_b = \mathrm{E}\left[\sum_{k=0}^{2\beta-1} s_c^2(kT_c)\right] = \mathrm{E}\left[\sum_{k=0}^{2\beta-1} c_k^2\right] \qquad (2.33)$$

and the variance of the discrete noise source is

$$\sigma_n^2 = W N_0'. \qquad (2.34)$$

[3] The quantity 2β is known as *spreading factor* in conventional spread-spectrum communication systems.

Substituting (2.33) and (2.34) into (2.32), we obtain

$$\zeta \equiv \frac{E_p}{N_0'} = \frac{E_b}{2\sigma_n^2} = \frac{E_b}{N_0} \tag{2.35}$$

where the last equality follows from the definition $N_0 = 2\sigma_n^2$. Thus, from (2.35), we observe that if we normalize the discrete noise source and model it as a Gaussian random variable with zero-mean and variance $N_0/2$ [Sushchik *et al.* (2000)], then ζ can be directly computed from E_b/N_0. For notational brevity, throughout this book, we use E_b/N_0 to denote the bit-energy-to-noise-power-spectral-density ratio, which has been the custom in much of the literature.

As a final remark before closing this chapter, the aforementioned discrete-time modelling approach is equally applicable to the analysis of conventional direct-sequence spread-spectrum systems where the binary symbols are spread by an aperiodic real chaotic sequence $\{c_k\}$ with a spreading factor of 2β.

3. Performance Analysis Methods for Coherent Chaos-Shift-Keying Systems

In this chapter we begin our formal study of chaos-based digital communication systems. The areas of investigation to be described here include the operation, analysis and performance evaluation of communication systems that employ chaotic signals for carrying digital information. As surveyed in the previous chapter, the most widely studied chaos-based digital communication systems are the so-called *chaos-shift-keying* (CSK) system [Dedieu *et al.* (1993)] and the *differential chaos-shift-keying* (DCSK) system [Kolumbán *et al.* (1996)]. In this chapter we focus our attention on the CSK system, in which digital symbols are encoded with different chaotic signals. In the binary case, the CSK transmitter essentially transmits two chaotic signals which are derived from two chaos generating sources, one at a time, according to the binary signal being sent. Clearly, the receiver is able to demodulate the transmitted signal if it can distinguish between the chaotic signal segments generated from the two chaos generators. Alternatively, the transmitted signal may be generated from one chaos generator in which a parameter is modulated according to the binary signal being sent. In this case, the receiver's job is to work out the parametric change by observing the transmitted signal.

In general, detection approaches for the CSK system can take either coherent or non-coherent forms [Kolumbán *et al.* (1998a); Kennedy and Kolumbán (2000)]. Coherent detection requires exact replicas of the chaotic signals used to carry the information be reproduced at the receiver, whereas non-coherent detection does not have such a requirement. In this chapter, we devote our effort to the study of the coherent CSK system, which employs a general correlator-type detection method. In particular we will develop analytical methods based on a discrete-time approach to evaluate the performance of the coherent CSK system under the assumption of an additive white Gaussian noise (AWGN) channel. A general *mixed analysis-simulation* approach for studying the noise performance of chaos-based digital communication systems will be discussed. Moreover, closed-form analytical expressions for the bit error rates will be derived for specific cases where certain conditions are met. Both single-user and multiple access systems will be considered.

Before embarking on the detailed analysis of the coherent CSK system, we should stress that coherent systems in general pose a practicality problem because of the lack of robust techniques for chaos synchronization which are needed for coherent detection. At the time of writing, this issue remains unresolved and there are strong evidence that non-coherent systems could be more realistic in terms of feasibility for practical implementation [Chen and Yao (2000); Kennedy and Kolumbán (2000a)]. Nonetheless, coherent systems are theoretically superior systems under the assumption of an AWGN channel. Thus, the study of coherent demodulation schemes is important as it can provide performance benchmarks for evaluating other chaos-based communication techniques under similar conditions.

3.1 Review of the Chaos-Shift-Keying (CSK) System

The CSK system was introduced by Parlitz *et al.* [1992] and Dedieu *et al.* [1993]. In its original form, the CSK system encodes binary symbols with two different chaotic attractors generated by two Chua's oscillators and its detection principle is based on the self-synchronizing property of the Chua's oscillators. Moreover, we may extend the idea to a general coherent communication system which encodes binary symbols with different chaotic reference signals and the detector recovers the binary symbols by correlating the received signal with the locally generated copies of the chaotic reference signals [Kolumbán (2000b)]. The rest of the chapter will be devoted to this type of communication systems.

As is customary in the analysis of communication systems, the baseband equivalent model shown in Fig. 3.1 is used in our initial examination of the coherent CSK system (see Sect. 2.6). Also, for simplicity, the system operation will be examined in a discrete-time fashion, i.e., all signals are treated as sequences of numbers. We denote the output sequences of the chaotic generators by $\{\hat{x}_k\}$ and $\{\check{x}_k\}$, respectively. We further denote the transmitted data by $\{\alpha_1, \alpha_2, \alpha_3, \ldots\}$, where $\alpha_l \in \{-1, +1\}$. Also, let 2β be the spreading factor, defined as the number of chaotic samples sent for each binary symbol. During the lth bit duration, i.e., for time $k = 2\beta(l-1)+1, 2\beta(l-1)+2, \ldots, 2\beta l$, the output of the transmitter, s_k, is

$$s_k = \begin{cases} \hat{x}_k & \text{if } \alpha_l = +1 \\ \check{x}_k & \text{if } \alpha_l = -1. \end{cases} \tag{3.1}$$

The transmitted signal, contaminated by additive white Gaussian noise, is presented to the receiver for decoding. At time k, the received signal, r_k, is given by

$$r_k = s_k + \xi_k \tag{3.2}$$

where ξ_k denotes the additive white Gaussian noise with zero mean and variance (power spectral density) $N_0/2$ (see Sect. 2.6).

As mentioned before, the coherent CSK system assumes the availability of exact synchronized copies of the chaotic signals at the receiver. We further

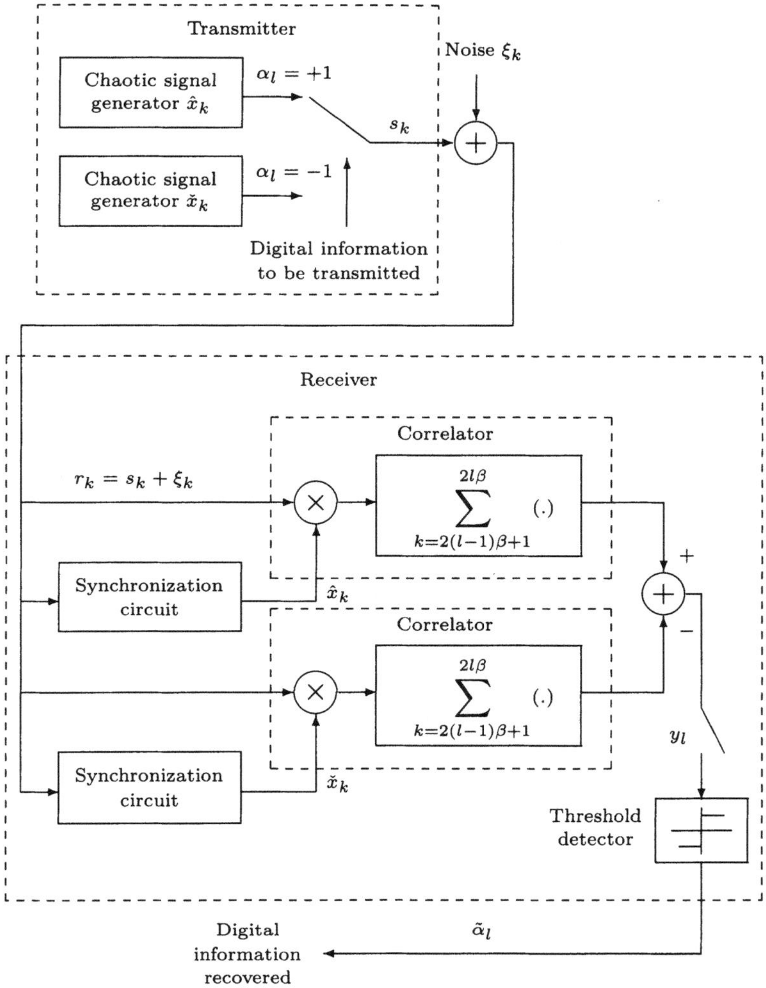

Fig. 3.1. A single-user coherent CSK communication system.

assume that the synchronization time is negligibly small compared to the bit period. Referring to Fig. 3.1, at the end of the lth bit duration, the input to the threshold detector is given by

$$y_l = \sum_{k=2(l-1)\beta+1}^{2l\beta} r_k \hat{x}_k - \sum_{k=2(l-1)\beta+1}^{2l\beta} r_k \check{x}_k. \tag{3.3}$$

The decoding algorithm is then completed by the following simple decision rule:

$$\tilde{\alpha}_l = \begin{cases} +1 & \text{if } y_l > 0 \\ -1 & \text{if } y_l \leq 0 \end{cases} \tag{3.4}$$

where y_l is the correlator output and $\tilde{\alpha}_l$ is the estimated symbol for the lth bit. In the following section, we will extend this system to a multiple access environment.

3.2 Analysis of the CSK System with Multiple Access

3.2.1 Transmitter Structure

We consider a multiple access CSK (MA-CSK) communication system with N users. A block diagram of the system is shown in Fig. 3.2. In the transmitter of the ith user, a pair of chaotic sequences, denoted by $\{\hat{x}_k^{(i)}\}$ and $\{\check{x}_k^{(i)}\}$, are generated by a chaotic map with different initial conditions. In this system, consistent with most practical systems, the mean value of each of the chaotic sequences is zero in order to avoid dissipating dc power. In other words,

$$\mathrm{E}[\hat{x}_k^{(i)}] = \mathrm{E}[\check{x}_k^{(i)}] = 0 \qquad i = 1, 2, \ldots, N \tag{3.5}$$

where $\mathrm{E}[\psi]$ denotes the mean value of ψ. Now let $\alpha_l^{(i)} \in \{-1, +1\}$ be the lth transmitted symbol for the ith user and assume that "-1" and "$+1$" occur with equal probabilities for all users. We again denote the spreading factor by 2β, which is defined as the number of chaotic samples used to transmit one binary symbol. During the lth bit duration, i.e., for time $k = 2(l-1)\beta + 1, 2(l-1)\beta + 2, \ldots, 2l\beta$, the output of the transmitter of the ith user, denoted by $s_k^{(i)}$, is

$$s_k^{(i)} = \begin{cases} \hat{x}_k^{(i)} & \text{if } \alpha_l^{(i)} = +1 \\ \check{x}_k^{(i)} & \text{if } \alpha_l^{(i)} = -1. \end{cases} \tag{3.6}$$

The overall transmitted signal of the multiple access system at time k, denoted by s_k, can be found by summing the signals of all users, i.e.,

$$s_k = \sum_{i=1}^{N} s_k^{(i)}. \tag{3.7}$$

3.2.2 Receiver Structure

The receiver structure of the jth user is depicted in Fig. 3.2. The transmitted signal is corrupted by additive noise, and the received signal, denoted by r_k, is thus given by

$$r_k = s_k + \xi_k \tag{3.8}$$

where ξ_k denotes the additive white Gaussian noise with zero mean and variance (power spectral density) $N_0/2$. Assuming that synchronized copies

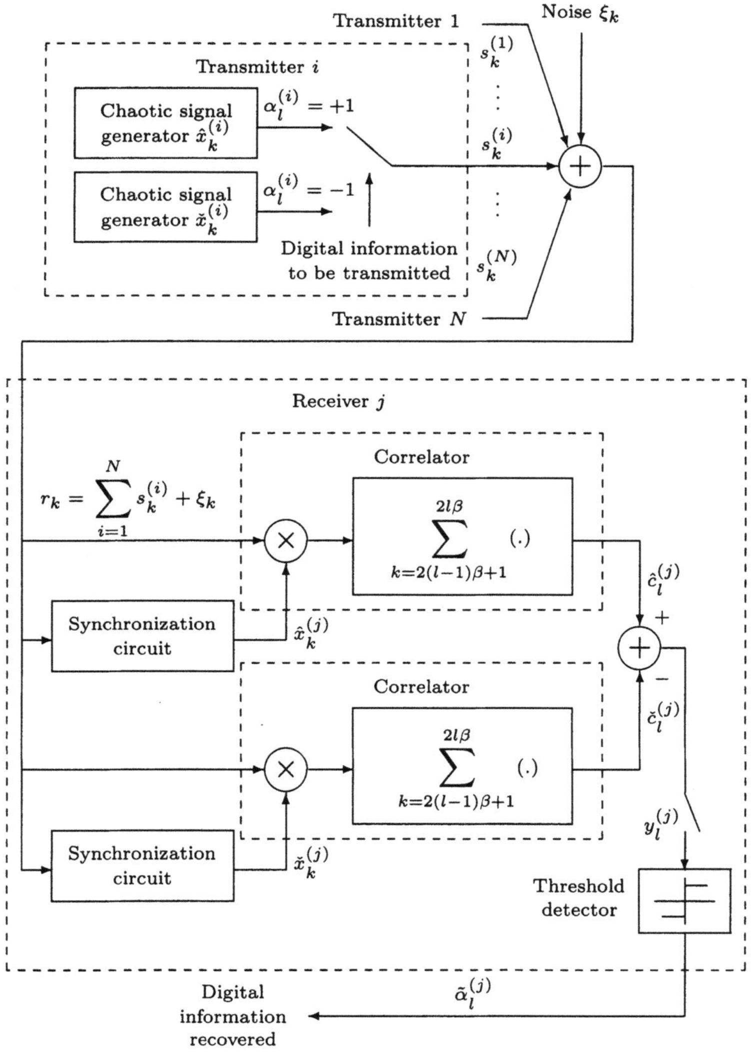

Fig. 3.2. A multiple access CSK communication system.

of the chaotic signals $\{\hat{x}_k^{(j)}\}$ and $\{\check{x}_k^{(j)}\}$ can be reproduced at the receiver, the detection essentially involves correlating the incoming signal with the locally regenerated chaotic signals and sampling the outputs of the correlators at the end of each symbol duration. The outputs of the correlators, as shown in Fig. 3.2, are given by

$$\hat{c}_l^{(j)} = \sum_{k=2(l-1)\beta+1}^{2l\beta} r_k \hat{x}_k^{(j)} \tag{3.9}$$

and

$$\check{c}_l^{(j)} = \sum_{k=2(l-1)\beta+1}^{2l\beta} r_k \check{x}_k^{(j)} \tag{3.10}$$

which are compared and sent to the threshold detector. The input to the detector at this time instant, denoted by $y_l^{(j)}$, is given by

$$y_l^{(j)} = \hat{c}_l^{(j)} - \check{c}_l^{(j)}. \tag{3.11}$$

The lth decoded symbol for the jth user, denoted by $\tilde{\alpha}_l^{(j)}$, is determined according to the following rule:

$$\tilde{\alpha}_l^{(j)} = \begin{cases} +1 & \text{if } y_l^{(j)} > 0 \\ -1 & \text{if } y_l^{(j)} \leq 0. \end{cases} \tag{3.12}$$

3.2.3 Derivation of Bit Error Rate

Without loss of generality, we consider the probability of error for the first symbol sent by the jth user. For brevity, the subscripts of the variables $\alpha_l^{(j)}, \tilde{\alpha}_l^{(j)}, \hat{c}_l^{(j)}, \check{c}_l^{(j)}$ and $y_l^{(j)}$ are omitted. The input to the threshold detector at the end of the first symbol duration, now denoted by $y^{(j)}$, is given by

$$\begin{aligned} y^{(j)} &= \hat{c}^{(j)} - \check{c}^{(j)} \\ &= \sum_{i=1}^{N}\sum_{k=1}^{2\beta} s_k^{(i)} \hat{x}_k^{(j)} - \sum_{i=1}^{N}\sum_{k=1}^{2\beta} s_k^{(i)} \check{x}_k^{(j)} + \sum_{k=1}^{2\beta} \xi_k \hat{x}_k^{(j)} - \sum_{k=1}^{2\beta} \xi_k \check{x}_k^{(j)}. \end{aligned} \tag{3.13}$$

If "+1" is transmitted for the jth user, i.e., $\alpha^{(j)} = +1$ and $s_k^{(j)} = \hat{x}_k^{(j)}$, (3.13) becomes

$$\begin{aligned} y^{(j)}|(\alpha^{(j)} = +1) &= \underbrace{V^{(j,j)} + W^{(j,j)}}_{\text{required signal}} + \underbrace{\sum_{i=1, i\neq j}^{N} V^{(i,j)} + \sum_{i=1, i\neq j}^{N} W^{(i,j)}}_{\text{inter-user interference}} \\ &\quad + \underbrace{X^{(j)} + Y^{(j)}}_{\text{noise}} \end{aligned} \tag{3.14}$$

where

$$V^{(i,j)} = \sum_{k=1}^{2\beta} s_k^{(i)} \hat{x}_k^{(j)} \tag{3.15}$$

$$W^{(i,j)} = -\sum_{k=1}^{2\beta} s_k^{(i)} \check{x}_k^{(j)} \tag{3.16}$$

$$X^{(j)} = \sum_{k=1}^{2\beta} \xi_k \hat{x}_k^{(j)} \tag{3.17}$$

$$Y^{(j)} = -\sum_{k=1}^{2\beta} \xi_k \check{x}_k^{(j)}. \tag{3.18}$$

Notice that the input to the detector consists of three components, namely, the required signal, inter-user interference and noise.

We assume that all signals (chaotic or noise) are stationary, i.e., the statistical properties of the signals are not dependent on time k. Using (3.5) and the fact that chaotic sequences from different generators are independent of one another, it can be shown that the mean value of $y^{(j)}|(\alpha^{(j)} = +1)$ equals

$$\begin{aligned} \mathrm{E}[y^{(j)}|(\alpha^{(j)} = +1)] &= \mathrm{E}[V^{(j,j)}] + \mathrm{E}[W^{(j,j)}] + \sum_{i=1, i\neq j}^{N} \mathrm{E}[V^{(i,j)}] \\ &\quad + \sum_{i=1, i\neq j}^{N} \mathrm{E}[W^{(i,j)}] + \mathrm{E}[X^{(j)}] + \mathrm{E}[Y^{(j)}] \\ &= \mathrm{E}[V^{(j,j)}] \\ &= 2\beta \mathrm{E}[(\hat{x}_k^{(j)})^2]. \end{aligned} \tag{3.19}$$

The variance of $y^{(j)}|(\alpha^{(j)} = +1)$ is found from

$$\begin{aligned} \mathrm{var}[y^{(j)}|(\alpha^{(j)} = +1)] &= \mathrm{var}[V^{(j,j)}] + \mathrm{var}[W^{(j,j)}] + \sum_{i=1, i\neq j}^{N} \mathrm{var}[V^{(i,j)}] \\ &\quad + \sum_{i=1, i\neq j}^{N} \mathrm{var}[W^{(i,j)}] + \mathrm{var}[X^{(j)}] + \mathrm{var}[Y^{(j)}] \\ &\quad + \sum_{\substack{C \\ C\neq D}} \sum_{D} \mathrm{cov}[C, D] \end{aligned} \tag{3.20}$$

where $C, D \in \{V^{(i,j)}, W^{(i,j)}, X^{(j)}, Y^{(j)}; i, j = 1, \ldots, N\}$, $\mathrm{var}[\psi]$ denotes the variance of ψ, and $\mathrm{cov}[C, D]$ represents the covariance of C and D defined as

$$\mathrm{cov}[C, D] = \mathrm{E}[CD] - \mathrm{E}[C]\mathrm{E}[D]. \tag{3.21}$$

All variance and covariance terms can be readily derived (see Appendix 3A). Also, it can be readily shown that all the covariance terms are zero . Thus, using the formulae (3.52)–(3.57), we may simplify (3.20) to

$$\mathrm{var}[y^{(j)}|(\alpha^{(j)} = +1)] = 2\beta \mathrm{var}[(\hat{x}_k^{(j)})^2] + \sum_{\substack{k=1 \\ k\neq m}}^{2\beta} \sum_{m=1}^{2\beta} \mathrm{cov}[(\hat{x}_k^{(j)})^2, (\hat{x}_m^{(j)})^2]$$

$$
\begin{aligned}
&+ 2\beta \mathrm{E}^2[(\hat{x}_k^{(j)})^2] + \sum_{\substack{k=1\\k\neq m}}^{2\beta} \sum_{m=1}^{2\beta} \mathrm{E}^2[\hat{x}_k^{(j)}\hat{x}_m^{(j)}] \\
&+ 4\beta \mathrm{E}[(\hat{x}_k^{(j)})^2] \sum_{\substack{i=1\\i\neq j}}^{N} \mathrm{E}[(\hat{x}_k^{(i)})^2] \\
&+ 2\sum_{\substack{i=1\\i\neq j}}^{N} \sum_{\substack{k=1\\k\neq m}}^{2\beta} \sum_{m=1}^{2\beta} \mathrm{E}[\hat{x}_k^{(i)}\hat{x}_m^{(i)}]\mathrm{E}[\hat{x}_k^{(j)}\hat{x}_m^{(j)}] \\
&+ 2\beta N_0 \mathrm{E}[(\hat{x}_k^{(j)})^2]. \qquad (3.22)
\end{aligned}
$$

Likewise, we can find $y^{(j)}|(\alpha^{(j)} = -1)$ and confirm that

$$\mathrm{E}[y^{(j)}|(\alpha^{(j)} = -1)] = -\mathrm{E}[y^{(j)}|(\alpha^{(j)} = +1)] \qquad (3.23)$$

and

$$\mathrm{var}[y^{(j)}|(\alpha^{(j)} = -1)] = \mathrm{var}[y^{(j)}|(\alpha^{(j)} = +1)]. \qquad (3.24)$$

Since $y^{(j)}|(\alpha^{(j)} = +1)$ and $y^{(j)}|(\alpha^{(j)} = -1)$ are the sum of a large number of random variables, we may assume that they are normally distributed. In the detection of the symbol, an error occurs if $y^{(j)} \leq 0 \mid (\alpha^{(j)} = +1)$ or $y^{(j)} > 0 \mid (\alpha^{(j)} = -1)$. The *bit error rate* (BER) for the jth user can thus be computed from

$$
\begin{aligned}
\mathrm{BER}^{(j)} &= \frac{1}{2}\mathrm{Prob}(y^{(j)} \leq 0 \mid (\alpha^{(j)} = +1)) + \frac{1}{2}\mathrm{Prob}(y^{(j)} > 0 \mid (\alpha^{(j)} = -1)) \\
&= \frac{1}{2}\mathrm{erfc}\left(\frac{\mathrm{E}[y^{(j)}|(\alpha^{(j)} = +1)]}{\sqrt{2\mathrm{var}[y^{(j)}|(\alpha^{(j)} = +1)]}}\right) \qquad (3.25)
\end{aligned}
$$

where the complementary error function, erfc(.), is defined as

$$\mathrm{erfc}(\psi) \equiv \frac{2}{\sqrt{\pi}} \int_{\psi}^{\infty} e^{-\lambda^2} d\lambda. \qquad (3.26)$$

3.2.3.1 Mixed Analysis-Simulation Technique. To find the BER, we first obtain the values of the terms (expected values, variances and covariances) on the right hand side of (3.19) and (3.22) by numerical simulations. Then, the values of $\mathrm{E}[y^{(j)}|(\alpha^{(j)} = +1)]$ and $\mathrm{var}[y^{(j)}|(\alpha^{(j)} = +1)]$ can be computed and substituted into (3.25) to obtain the BER. Since both numerical simulation and analytical method are involved in obtaining the BER, we refer to this method as *mixed analysis-simulation* (MAS) technique.

3.2.3.2 Analytical BERs. When certain specific conditions hold, closed-form analytical expressions for the BERs may be derived. First of all, we assume that all users transmit with equal *average power*, i.e., the mean squared values of all chaotic signals are the same. Writing the average power, P_s, as

$$\begin{aligned} P_s &= \mathrm{E}[(\hat{x}_k^{(1)})^2] = \mathrm{E}[(\hat{x}_k^{(2)})^2] = \cdots = \mathrm{E}[(\hat{x}_k^{(N)})^2] \\ &= \mathrm{E}[(\check{x}_k^{(1)})^2] = \mathrm{E}[(\check{x}_k^{(2)})^2] = \cdots = \mathrm{E}[(\check{x}_k^{(N)})^2], \end{aligned} \tag{3.27}$$

we may simplify (3.19) and (3.22) to

$$\mathrm{E}[y^{(j)}|(\alpha^{(j)} = +1)] = 2\beta P_s \tag{3.28}$$

and

$$\begin{aligned} \mathrm{var}[y^{(j)}|(\alpha^{(j)} = +1)] &= 2\beta \mathrm{var}[(\hat{x}_k^{(j)})^2] + 2\beta(2N-1)P_s^2 + 2\beta N_0 P_s \\ &+ \sum_{\substack{k=1 \\ k\neq m}}^{2\beta} \sum_{m=1}^{2\beta} \mathrm{cov}[(\hat{x}_k^{(j)})^2, (\hat{x}_m^{(j)})^2] + \sum_{\substack{k=1 \\ k\neq m}}^{2\beta} \sum_{m=1}^{2\beta} \mathrm{E}^2[\hat{x}_k^{(j)} \hat{x}_m^{(j)}] \\ &+ 2\sum_{\substack{i=1 \\ i\neq j}}^{N} \sum_{\substack{k=1 \\ k\neq m}}^{2\beta} \sum_{m=1}^{2\beta} \mathrm{E}[\hat{x}_k^{(i)} \hat{x}_m^{(i)}] \mathrm{E}[\hat{x}_k^{(j)} \hat{x}_m^{(j)}]. \end{aligned} \tag{3.29}$$

Special Case I: Chaotic sequences with particular statistical properties

The three summation terms on the right hand side of (3.29) can be eliminated if the chaotic sequence generated by the jth user, $\{\hat{x}_k^{(j)}\}$, satisfies the following conditions:

$$\mathrm{cov}[(\hat{x}_k^{(j)})^2, (\hat{x}_m^{(j)})^2] = 0 \quad k \neq m \tag{3.30}$$

$$\mathrm{E}[\hat{x}_k^{(j)} \hat{x}_m^{(j)}] = 0 \quad k \neq m. \tag{3.31}$$

These conditions essentially require that both the *autovariance* of $\{(\hat{x}_k^{(j)})^2\}$ and the *autocorrelation* of $\{\hat{x}_k^{(j)}\}$ vanish. If these conditions hold, the variance of $y^{(j)}|(\alpha^{(j)} = +1)$ reduces to

$$\mathrm{var}[y^{(j)}|(\alpha^{(j)} = +1)] = 2\beta \mathrm{var}[(\hat{x}_k^{(j)})^2] + 2\beta(2N-1)P_s^2 + 2\beta N_0 P_s. \tag{3.32}$$

For brevity of notations, we define

$$\Psi^{(j)} = \frac{\mathrm{var}[(\hat{x}_k^{(j)})^2]}{P_s^2} = \frac{\mathrm{var}[(\hat{x}_k^{(j)})^2]}{\mathrm{E}^2[(\hat{x}_k^{(j)})^2]}. \tag{3.33}$$

Suppose the chaotic sequence $\{\hat{x}_k^{(j)}\}$ is multiplied by the factor ν before transmission. Then, the average power of the signal becomes $\nu^2 P_s$. However, the value of $\Psi^{(j)}$ remains unchanged because

$$\Psi^{(j)} = \frac{\mathrm{var}[(\nu\hat{x}_k^{(j)})^2]}{\mathrm{E}^2[(\nu\hat{x}_k^{(j)})^2]} = \frac{\mathrm{var}[\nu^2(\hat{x}_k^{(j)})^2]}{\mathrm{E}^2[\nu^2(\hat{x}_k^{(j)})^2]} = \frac{\nu^4\mathrm{var}[(\hat{x}_k^{(j)})^2]}{\nu^4\mathrm{E}^2[(\hat{x}_k^{(j)})^2]} = \frac{\mathrm{var}[(\hat{x}_k^{(j)})^2]}{\mathrm{E}^2[(\hat{x}_k^{(j)})^2]}. \tag{3.34}$$

In other words, the value $\Psi^{(j)}$ is a constant, dependent upon only the type of the chaotic sequence but not the average power of the sequence. Now, we rewrite (3.32) as

$$\mathrm{var}[y^{(j)}|(\alpha^{(j)} = +1)] = 2\beta\Psi^{(j)}P_s^2 + 2\beta(2N-1)P_s^2 + 2\beta N_0 P_s \tag{3.35}$$

and the analytical BER for the jth user can be expressed as

$$\begin{aligned}
\mathrm{BER}_{\text{CSK-I}}^{(j)} &= \frac{1}{2}\mathrm{erfc}\left(\left[\frac{\Psi^{(j)}}{\beta} + \frac{(2N-1)}{\beta} + \frac{N_0}{\beta P_s}\right]^{-\frac{1}{2}}\right) &(3.36)\\
&= \frac{1}{2}\mathrm{erfc}\left(\left[\frac{\Psi^{(j)}}{\beta} + \frac{(2N-1)}{\beta} + 2\left(\frac{E_b}{N_0}\right)^{-1}\right]^{-\frac{1}{2}}\right) &(3.37)
\end{aligned}$$

where E_b represents the *average bit energy* and is given by

$$E_b = 2\beta P_s. \tag{3.38}$$

To see how the various parameters affect the BER, it is convenient to define

$$\begin{aligned}
\theta_{\text{CSK-I}} &= \underbrace{\frac{\Psi^{(j)}}{\beta}}_{\text{variation of } (\hat{x}_k^{(j)})^2} + \underbrace{\frac{(2N-1)}{\beta}}_{\text{“interfering” signals acting on the correlators}} + \underbrace{2\left(\frac{E_b}{N_0}\right)^{-1}}_{\text{noise component}} &(3.39)\\
&= \frac{\Psi^{(j)}}{\beta} + \frac{(2N-1)}{\beta} + 2\left(\frac{E_b}{N_0}\right)^{-1}. &(3.40)
\end{aligned}$$

Since the complementary error function, erfc(.), is a monotonically decreasing function, we observe that $\mathrm{BER}_{\text{CSK-I}}^{(j)}$ increases with $\theta_{\text{CSK-I}}$. The first term on the right hand side of (3.39) arises from the variation of $(\hat{x}_k^{(j)})^2$, and it can be reduced by choosing a chaotic sequence with smaller $\mathrm{var}[(\hat{x}_k^{(j)})^2]$ for a fixed P_s. The second term is due to the “interfering” signals acting on the two correlators of the user. Among the N incoming signal components, only one is the “required” signal for one of the two correlators. For the other correlator, all the N components are regarded as interferers. Thus, there is a total of $(2N-1)$ “interfering” signals, giving rise to the factor $(2N-1)$ in the second term. The third term in (3.39) is due to noise and it can be reduced by increasing the bit energy of all users. Therefore, for a fixed E_b/N_0 value, we can improve the BER for the jth user (reduce the parameter $\theta_{\text{CSK-I}}$) by making one or more of the following adjustments.

1. Increase the spreading factor 2β.

2. Minimize the variance of $\{(\hat{x}_k^{(j)})^2\}$ for a fixed P_s. In the optimal case where $\mathrm{var}[(\hat{x}_k^{(j)})^2] = 0$, binary sequences will be produced by the chaotic generators.
3. Reduce the number of users N.

Special Case II: All users use the same chaotic map to generate sequences with particular statistical properties

Suppose all users employ the same chaos generating map, and each uses a different initial condition. The statistical properties of $\{\hat{x}_k^{(i)}\}$ will be identical for all $i = 1, 2, \ldots, N$. Also, if the autovariance of $\{(\hat{x}_k^{(i)})^2\}$ and the autocorrelation of $\{\hat{x}_k^{(i)}\}$ vanish for all $i = 1, 2, \ldots, N$, (3.29) becomes

$$\mathrm{var}[y^{(j)} | (\alpha^{(j)} = +1)] = 2\beta \Psi P_s^2 + 2\beta(2N-1)P_s^2 + 2\beta N_0 P_s \tag{3.41}$$

where

$$\Psi = \frac{\mathrm{var}[(\hat{x}_k^{(1)})^2]}{P_s^2} = \frac{\mathrm{var}[(\hat{x}_k^{(2)})^2]}{P_s^2} = \cdots = \frac{\mathrm{var}[(\hat{x}_k^{(N)})^2]}{P_s^2}. \tag{3.42}$$

The BER for each user is thus the same and is given by

$$\mathrm{BER}_{\text{CSK-II}} = \frac{1}{2}\mathrm{erfc}\left(\left[\frac{\Psi}{\beta} + \frac{(2N-1)}{\beta} + \frac{N_0}{\beta P_s}\right]^{-\frac{1}{2}}\right) \tag{3.43}$$

$$= \frac{1}{2}\mathrm{erfc}\left(\left[\frac{\Psi}{\beta} + \frac{(2N-1)}{\beta} + 2\left(\frac{E_b}{N_0}\right)^{-1}\right]^{-\frac{1}{2}}\right) \tag{3.44}$$

where E_b is defined as in (3.38). As in the previous case, for a fixed E_b/N_0, the BER for each user can be improved by making one or more of the following adjustments.

1. Increase the spreading factor 2β.
2. Minimize the variance of $\{\hat{x}_k^2\}$ for a fixed P_s.
3. Reduce the number of users N.

3.3 Simulations and Evaluation

The following maps will be used in the simulation studies.

1. Logistic map

$$x_{k+1} = 1 - 2x_k^2 \tag{3.45}$$

2. Cubic map

$$x_{k+1} = 4x_k^3 - 3x_k \tag{3.46}$$

Table 3.1. Statistical properties of the normalized chaotic sequences.

	Logistic map	Cubic map	Skew tent map	Hénon map	Bernoulli-shift map
$\mathrm{E}[x_k]$	0	0	0	0	0
$\mathrm{E}[x_k^2]$	1	1	1	1	1
$\mathrm{E}[x_k^3]$	0	0	0	−0.4950	0
$\mathrm{var}[x_k]$	1	1	1	1	1
$\mathrm{var}[x_k^2]$	0.5	0.5	0.8	1.1216	0.9874

3. Skew tent map

$$x_{k+1} = \begin{cases} \dfrac{x_k}{0.6} & \text{when } 0 \leq x_k \leq 0.6 \\ \dfrac{1-x_k}{0.4} & \text{when } 0.6 < x_k \leq 1 \end{cases} \tag{3.47}$$

4. Hénon map

$$\begin{aligned} x_{k+1} &= 1 + y_k - 1.4x_k^2 \\ y_{k+1} &= 0.3x_k \end{aligned} \tag{3.48}$$

5. Bernoulli-shift map

$$x_{k+1} = \begin{cases} 1.2x_k + 1 & \text{when } x_k < 0 \\ 1.2x_k - 1 & \text{when } x_k > 0 \end{cases} \tag{3.49}$$

In the simulations, the chaotic sequences generated by (3.45) to (3.49) are normalized such that their mean values are all zero and their mean squared values are unity, i.e., $\mathrm{E}[x_k] = 0$ and $\mathrm{E}[x_k^2] = 1$. The statistical properties of the normalized sequences are shown in Table 3.1. It is readily shown (see Appendix 3B) that the chaotic sequences generated from the logistic map and the cubic map satisfy the statistical properties assumed in Sect. 3.2.3.2, i.e.,

$$\mathrm{cov}[x_k^2, x_m^2] = \mathrm{E}[x_k^2 x_m^2] - \mathrm{E}[x_k^2]\mathrm{E}[x_m^2] = 0 \quad k \neq m \tag{3.50}$$

$$\mathrm{E}[x_k x_m] = 0 \quad k \neq m. \tag{3.51}$$

In the following we present three types of results. The first type is derived from the afore-discussed mixed analysis-simulation (MAS) technique. The second type of results is based on the analytical expressions derived in Sect. 3.2.3.2. The third type of results is based on brute-force (BF) simulations using the empirical models without making any simplifying assumptions, and hence can present true system performance for comparison and verification purposes.

3.3.1 Users Using Distinct Chaotic Maps

We consider a 5-user MA-CSK system in which different maps are used by different users:

User 1: Logistic map
User 2: Cubic map
User 3: Skew tent map
User 4: Hénon map
User 5: Bernoulli-shift map

The BERs versus E_b/N_0 for all users are plotted in Figs. 3.3 and 3.4 for $2\beta = 100$ and 1000, respectively. Note that the mixed analysis-simulation (MAS) results are obtained from (3.19), (3.22) and (3.25). Also, in Figs. 3.3 (a) and 3.4 (a), the analytical BERs for Users 1 and 2 are obtained from (3.37). In all cases, the brute-force (BF) simulated results are shown for the sake of comparison and verification. It should be noted that the BERs are different for users with different chaotic sequences. From Figs. 3.3 and 3.4, it is readily seen that BER performance improves as the spreading factor increases.

3.3.2 Same Chaotic Map Used by All Users

Next, we simulate a coherent MA-CSK system in which all chaotic signals are generated by the same map but with different initial conditions. When the logistic map or the cubic map is used, the analytical BER for all users is given by (3.44).

In the first simulation, the cubic map (3.46) is used and the spreading factor is 100. In Fig. 3.5, the BERs are plotted against E_b/N_0 for a 3-user system. Furthermore, in Fig. 3.6, the BERs are plotted against the number of users for $E_b/N_0 = 10$ dB. As clearly shown, when the number of users increases, the BER increases (deteriorates) because of the increased level of inter-user interference. The results also indicate that the analytical method can provide a good estimation of the performance of the actual system.

Finally, in Fig. 3.7, we present the brute-force simulated BERs against E_b/N_0 for the cases corresponding to different choices of chaotic maps. It is found that the chaotic sequences generated by the Bernoulli-shift map produce a higher BER, and that the BERs for the users using the cubic map and the logistic map are the same. This observation can be easily explained in terms of the value of $\text{var}[x_k^2]$ which is larger for the case of the Bernoulli-shift map (see Table 3.1).

In concluding this chapter, we re-emphasize that in evaluating the performance of the coherent multiple access CSK system, perfect synchronization of chaotic signals at the receiving end has been assumed. Although robust synchronization techniques are not yet available, the study of coherent MA-CSK schemes can provide benchmark performance for comparison with other chaotic modulation schemes.

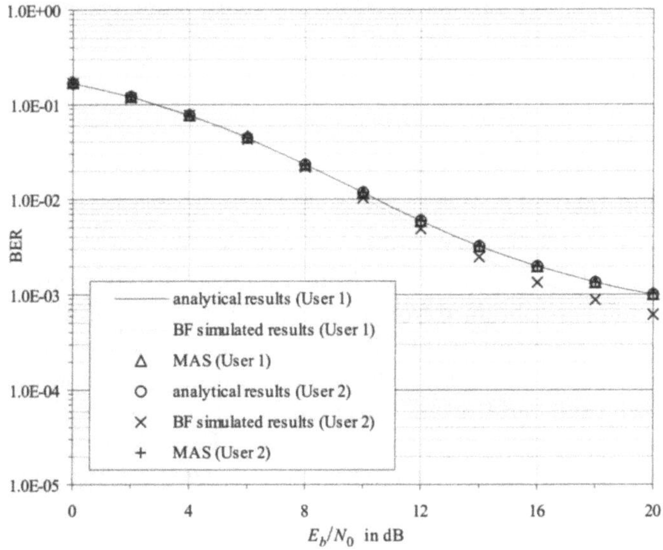

(a)

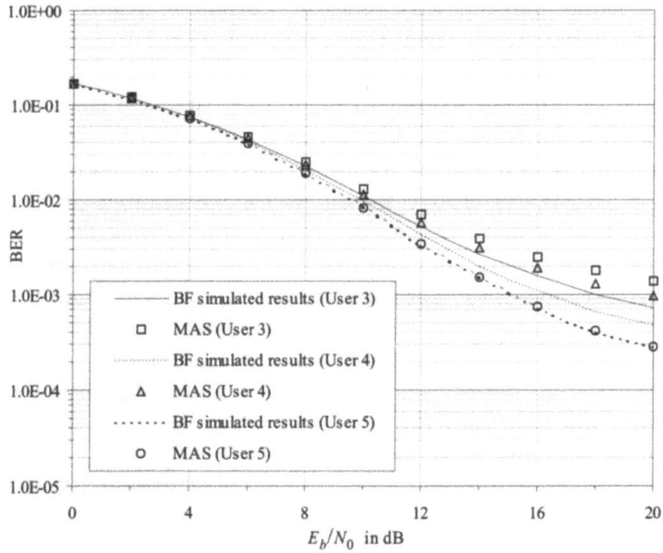

(b)

Fig. 3.3. BER versus E_b/N_0 in a 5-user CSK system. Analytical, mixed analysis-simulation (MAS) and brute-force (BF) simulated results are shown. Logistic map for User 1, cubic map for User 2, skew tent map for User 3, Hénon map for User 4, Bernoulli-shift map for User 5. $2\beta = 100$. (a) Users 1 and 2; (b) Users 3 to 5.

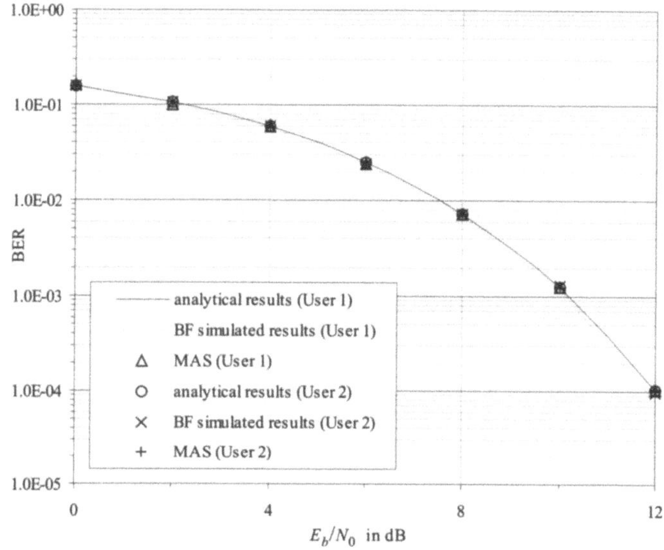

(a)

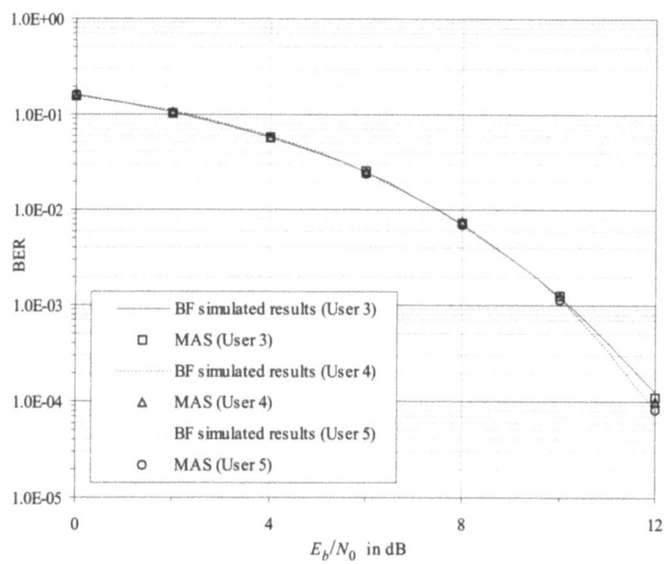

(b)

Fig. 3.4. BER versus E_b/N_0 in a 5-user CSK system. Analytical, mixed analysis-simulation (MAS) and brute-force (BF) simulated results are shown. Logistic map for User 1, cubic map for User 2, skew tent map for User 3, Hénon map for User 4, Bernoulli-shift map for User 5. $2\beta = 1000$. (a) Users 1 and 2; (b) Users 3 to 5.

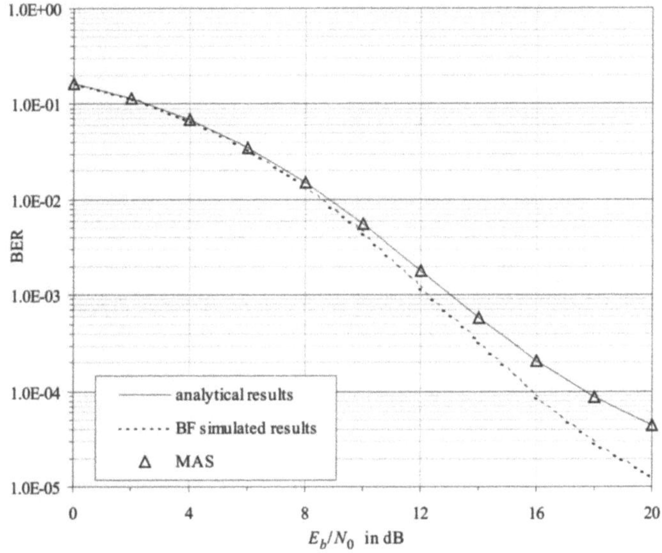

Fig. 3.5. BER versus E_b/N_0 in a 3-user MA-CSK system. Analytical, brute-force (BF) simulation and mixed analysis-simulation (MAS) results are shown for comparison. Cubic map is used and $2\beta = 100$.

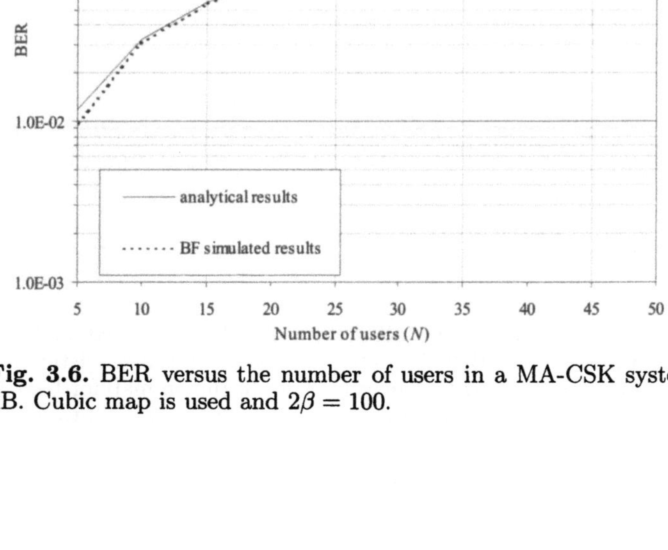

Fig. 3.6. BER versus the number of users in a MA-CSK system at $E_b/N_0 = 10$ dB. Cubic map is used and $2\beta = 100$.

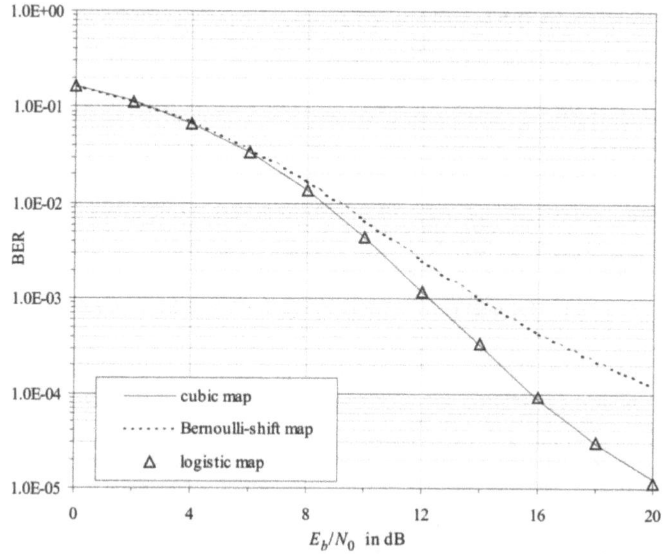

Fig. 3.7. Brute-force simulated BER versus E_b/N_0 in a 3-user MA-CSK system. Cubic map, Bernoulli-shift map and logistic map are used. $2\beta = 100$.

Appendix 3A: Derivation of variances relevant to the analysis of multiple access CSK system

All symbols are defined as in Sect. 3.2.

1. The derivation of the variance of $V^{(i,j)}$ is as follows. For $i \neq j$,

$$
\begin{aligned}
\mathrm{var}[V^{(i,j)}] &= \mathrm{var}\left[\sum_{k=1}^{2\beta} s_k^{(i)}\hat{x}_k^{(j)}\right] \\
&= \sum_{k=1}^{2\beta}\mathrm{var}[s_k^{(i)}\hat{x}_k^{(j)}] + \sum_{\substack{k=1\\k\neq m}}^{2\beta}\sum_{m=1}^{2\beta}\mathrm{cov}[s_k^{(i)}\hat{x}_k^{(j)}, s_m^{(i)}\hat{x}_m^{(j)}] \\
&= \sum_{k=1}^{2\beta}\left[\mathrm{var}[s_k^{(i)}]\mathrm{var}[\hat{x}_k^{(j)}] + \mathrm{var}[s_k^{(i)}]\mathrm{E}^2[\hat{x}_k^{(j)}] + \mathrm{E}^2[s_k^{(i)}]\mathrm{var}[\hat{x}_k^{(j)}]\right] \\
&\quad + \sum_{\substack{k=1\\k\neq m}}^{2\beta}\sum_{m=1}^{2\beta}\left[\mathrm{E}[s_k^{(i)}\hat{x}_k^{(j)}s_m^{(i)}\hat{x}_m^{(j)}] - \mathrm{E}[s_k^{(i)}\hat{x}_k^{(j)}]\mathrm{E}[s_m^{(i)}\hat{x}_m^{(j)}]\right] \\
&= \sum_{k=1}^{2\beta}\mathrm{var}[s_k^{(i)}]\mathrm{var}[\hat{x}_k^{(j)}] + \sum_{\substack{k=1\\k\neq m}}^{2\beta}\sum_{m=1}^{2\beta}\mathrm{E}[s_k^{(i)}s_m^{(i)}]\mathrm{E}[\hat{x}_k^{(j)}\hat{x}_m^{(j)}] \\
&= 2\beta\mathrm{E}[(s_k^{(i)})^2]\mathrm{E}[(\hat{x}_k^{(j)})^2] + \sum_{\substack{k=1\\k\neq m}}^{2\beta}\sum_{m=1}^{2\beta}\mathrm{E}[s_k^{(i)}s_m^{(i)}]\mathrm{E}[\hat{x}_k^{(j)}\hat{x}_m^{(j)}] \\
&= 2\beta\mathrm{E}[(\hat{x}_k^{(i)})^2]\mathrm{E}[(\hat{x}_k^{(j)})^2] + \sum_{\substack{k=1\\k\neq m}}^{2\beta}\sum_{m=1}^{2\beta}\mathrm{E}[\hat{x}_k^{(i)}\hat{x}_m^{(i)}]\mathrm{E}[\hat{x}_k^{(j)}\hat{x}_m^{(j)}]. \qquad (3.52)
\end{aligned}
$$

The last equality holds because the statistical properties of the sequences $\{\hat{x}_k^{(i)}\}$ and $\{\breve{x}_k^{(i)}\}$ are identical.

2. For $i = j$, the derivation of the variance of $V^{(i,j)}$ is as follows.

$$
\begin{aligned}
\mathrm{var}[V^{(j,j)}] &= \mathrm{var}\left[\sum_{k=1}^{2\beta}(\hat{x}_k^{(j)})^2\right] \\
&= \sum_{k=1}^{2\beta}\mathrm{var}[(\hat{x}_k^{(j)})^2] + \sum_{\substack{k=1\\k\neq m}}^{2\beta}\sum_{m=1}^{2\beta}\mathrm{cov}[(\hat{x}_k^{(j)})^2, (\hat{x}_m^{(j)})^2] \\
&= 2\beta\mathrm{var}[(\hat{x}_k^{(j)})^2] + \sum_{\substack{k=1\\k\neq m}}^{2\beta}\sum_{m=1}^{2\beta}\mathrm{cov}[(\hat{x}_k^{(j)})^2, (\hat{x}_m^{(j)})^2] \qquad (3.53)
\end{aligned}
$$

3. For $i \neq j$, the derivation of $\mathrm{var}[W^{(i,j)}]$ is similar to that of $\mathrm{var}[V^{(i,j)}]$, and it is readily shown that

$$\mathrm{var}[W^{(i,j)}] = 2\beta \mathrm{E}[(\hat{x}_k^{(i)})^2]\mathrm{E}[(\hat{x}_k^{(j)})^2] + \sum_{\substack{k=1 \\ k\neq m}}^{2\beta} \sum_{m=1}^{2\beta} \mathrm{E}[\hat{x}_k^{(i)}\hat{x}_m^{(i)}]\mathrm{E}[\hat{x}_k^{(j)}\hat{x}_m^{(j)}]. \tag{3.54}$$

4. For $i = j$, the derivation of $\mathrm{var}[W^{(i,j)}]$ is as follows.

$$\begin{aligned}
\mathrm{var}[W^{(j,j)}] &= \mathrm{var}\left[\sum_{k=1}^{2\beta} \hat{x}_k^{(j)}\check{x}_k^{(j)}\right] \\
&= \sum_{k=1}^{2\beta} \mathrm{var}[\hat{x}_k^{(j)}\check{x}_k^{(j)}] + \sum_{\substack{k=1 \\ k\neq m}}^{2\beta} \sum_{m=1}^{2\beta} \mathrm{cov}[\hat{x}_k^{(j)}\check{x}_k^{(j)}, \hat{x}_m^{(j)}\check{x}_m^{(j)}] \\
&= \sum_{k=1}^{2\beta} \left[\mathrm{var}[\hat{x}_k^{(j)}]\mathrm{var}[\check{x}_k^{(j)}] + \mathrm{var}[\hat{x}_k^{(j)}]\mathrm{E}^2[\check{x}_k^{(j)}] + \mathrm{E}^2[\hat{x}_k^{(j)}]\mathrm{var}[\check{x}_k^{(j)}]\right] \\
&\quad + \sum_{\substack{k=1 \\ k\neq m}}^{2\beta} \sum_{m=1}^{2\beta} \left[\mathrm{E}[\hat{x}_k^{(j)}\check{x}_k^{(j)}\hat{x}_m^{(j)}\check{x}_m^{(j)}] - \mathrm{E}[\hat{x}_k^{(j)}\check{x}_k^{(j)}]\mathrm{E}[\hat{x}_m^{(j)}\check{x}_m^{(j)}]\right] \\
&= \sum_{k=1}^{2\beta} \mathrm{var}[\hat{x}_k^{(j)}]\mathrm{var}[\check{x}_k^{(j)}] + \sum_{\substack{k=1 \\ k\neq m}}^{2\beta} \sum_{m=1}^{2\beta} \mathrm{E}[\hat{x}_k^{(j)}\hat{x}_m^{(j)}]\mathrm{E}[\check{x}_k^{(j)}\check{x}_m^{(j)}] \\
&= 2\beta \mathrm{E}[(\hat{x}_k^{(j)})^2]\mathrm{E}[(\check{x}_k^{(j)})^2] + \sum_{\substack{k=1 \\ k\neq m}}^{2\beta} \sum_{m=1}^{2\beta} \mathrm{E}[\hat{x}_k^{(j)}\hat{x}_m^{(j)}]\mathrm{E}[\check{x}_k^{(j)}\check{x}_m^{(j)}] \\
&= 2\beta \mathrm{E}^2[(\hat{x}_k^{(j)})^2] + \sum_{\substack{k=1 \\ k\neq m}}^{2\beta} \sum_{m=1}^{2\beta} \mathrm{E}^2[\hat{x}_k^{(j)}\hat{x}_m^{(j)}] \qquad (3.55)
\end{aligned}$$

5. The derivation of the variance of $X^{(j)}$ is as follows.

$$\begin{aligned}
\mathrm{var}[X^{(j)}] &= \mathrm{var}\left[\sum_{k=1}^{2\beta} \xi_k \hat{x}_k^{(j)}\right] \\
&= \sum_{k=1}^{2\beta} \mathrm{var}[\xi_k \hat{x}_k^{(j)}] + \sum_{\substack{k=1 \\ k\neq m}}^{2\beta} \sum_{m=1}^{2\beta} \mathrm{cov}[\xi_k \hat{x}_k^{(j)}, \xi_l \hat{x}_m^{(j)}] \\
&= \sum_{k=1}^{2\beta} \left[\mathrm{var}[\xi_k]\mathrm{var}[\hat{x}_k^{(j)}] + \mathrm{var}[\xi_k]\mathrm{E}^2[\hat{x}_k^{(j)}] + \mathrm{E}^2[\xi_k]\mathrm{var}[\hat{x}_k^{(j)}]\right]
\end{aligned}$$

$$
\begin{aligned}
&+\sum_{\substack{k=1\\k\neq m}}^{2\beta}\sum_{m=1}^{2\beta}\left[\mathrm{E}[\xi_k\hat{x}_k^{(j)}\xi_m\hat{x}_m^{(j)}]-\mathrm{E}[\xi_k\hat{x}_k^{(j)}]\mathrm{E}[\xi_m\hat{x}_m^{(j)}]\right]\\
&= \sum_{k=1}^{2\beta}\mathrm{var}[\xi_k]\mathrm{var}[\hat{x}_k^{(j)}]+\sum_{\substack{k=1\\k\neq m}}^{2\beta}\sum_{m=1}^{2\beta}\mathrm{E}[\xi_k\xi_m]\mathrm{E}[\hat{x}_k^{(j)}\hat{x}_m^{(j)}]\\
&= 2\beta\mathrm{var}[\xi_k]\mathrm{var}[\hat{x}_k^{(j)}]\\
&= \beta N_0\mathrm{E}[(\hat{x}_k^{(j)})^2]
\end{aligned}
\tag{3.56}
$$

6. The derivation of $\mathrm{var}[Y^{(j)}]$ is similar to that of $\mathrm{var}[X^{(j)}]$, and it is readily shown that

$$
\mathrm{var}[Y^{(j)}]=\beta N_0\mathrm{E}[(\hat{x}_k^{(j)})^2]. \tag{3.57}
$$

Appendix 3B: Derivation of the statistical properties for the chaotic sequences generated by the logistic map and the cubic map

Suppose $\{x_k\}$ is a chaotic sequence generated by the logistic map $x_{k+1} = g(x_k) = 1 - 2x_k^2$. The invariant probability density function of x_k, denoted by $\rho(x)$, is given by [Kohda and Tsuneda (1994)]

$$
\rho(x)=\begin{cases}\dfrac{1}{\pi\sqrt{1-x^2}} & \text{if } |x|<1\\ 0 & \text{otherwise.}\end{cases} \tag{3.58}
$$

Since $\rho(x)$ is an even function, the mean value of x_k is

$$
\mathrm{E}[x_k]=\int_{-\infty}^{\infty}x\rho(x)dx=\int_{-1}^{1}x\rho(x)dx=0. \tag{3.59}
$$

Define

$$
\begin{aligned}
g^{(1)}(x) &= g(x)\\
g^{(2)}(x) &= g(g^{(1)}(x))\\
\vdots &= \vdots\\
g^{(n)}(x) &= g(g^{(n-1)}(x)).
\end{aligned}
\tag{3.60}
$$

Using the formula $1-2\cos^2 v=-\cos 2v$, we have

$$
\begin{aligned}
g(\cos\phi) &= 1-2\cos^2\phi=-\cos 2\phi\\
g^{(2)}(\cos\phi) &= g(g(\cos\phi))=g(-\cos 2\phi)=-\cos(2^2\phi)\\
g^{(3)}(\cos\phi) &= g(g^{(2)}(\cos\phi))=g(-\cos(2^2\phi))=-\cos(2^3\phi)\\
\vdots &= \vdots\\
g^{(n)}(\cos\phi) &= g(g^{(n-1)}(\cos\phi))=g(-\cos(2^{n-1}\phi))=-\cos(2^n\phi).
\end{aligned}
\tag{3.61}
$$

Derivation of $\mathrm{cov}[x_k^2, x_m^2]$ for logistic map: The autovariance of $\{x_k^2\}$ is given by

$$\mathrm{cov}[x_k^2, x_m^2] = \mathrm{E}[x_k^2 x_m^2] - \mathrm{E}[x_k^2]\mathrm{E}[x_m^2]. \tag{3.62}$$

We consider the case where $k \neq m$. Without loss of generality, assume $m = k + n$ for some positive integer n.

$$\begin{aligned} \mathrm{E}[x_k^2 x_m^2] &= \int_{-\infty}^{\infty} \rho(x)\; x^2\; (g^{(n)}(x))^2 dx \\ &= \int_{-1}^{1} \frac{1}{\pi\sqrt{1-x^2}}\; x^2\; (g^{(n)}(x))^2 dx. \end{aligned} \tag{3.63}$$

Making the substitution $x = \cos\phi$, (3.63) becomes

$$\begin{aligned} \mathrm{E}[x_k^2 x_m^2] &= \int_{\pi}^{0} \frac{1}{\pi \sin\phi} \cos^2\phi\;\; (g^{(n)}(\cos\phi))^2\; (-\sin\phi)\; d\phi \\ &= \frac{1}{\pi}\int_0^{\pi} \cos^2\phi\;\; (g^{(n)}(\cos\phi))^2\; d\phi. \end{aligned} \tag{3.64}$$

Substituting (3.61) into (3.64), we obtain

$$\begin{aligned} \mathrm{E}[x_k^2 x_m^2] &= \frac{1}{\pi}\int_0^{\pi} \cos^2\phi\; \cos^2(2^n\phi)\; d\phi \\ &= \frac{1}{\pi}\int_0^{\pi} \left(\frac{1+\cos 2\phi}{2}\right)\left(\frac{1+\cos(2^{n+1}\phi)}{2}\right) d\phi \\ &= \frac{1}{4\pi}\int_0^{\pi} \Bigg(1 + \cos 2\phi + \cos(2^{n+1}\phi) + \frac{1}{2}\cos((2^{n+1}+2)\phi) \\ &\qquad\qquad + \frac{1}{2}\cos((2^{n+1}-2)\phi)\Bigg)\; d\phi \\ &= \frac{1}{4\pi}\Bigg[\phi + \frac{1}{2}\sin 2\phi + \frac{1}{2^{n+1}}\sin(2^{n+1}\phi) \\ &\quad + \frac{1}{2^{n+2}+4}\sin((2^{n+1}+2)\phi) + \frac{1}{2^{n+2}-4}\sin((2^{n+1}-2)\phi)\Bigg]_0^{\pi} \\ &= \frac{1}{4}. \end{aligned} \tag{3.65}$$

Simliarly it is readily shown that

$$\mathrm{E}[x_k^2] = \mathrm{E}[x_m^2] = \int_{-\infty}^{\infty} \rho(x) x^2 dx = \int_{-1}^{1} \frac{1}{\pi\sqrt{1-x^2}}\; x^2 dx = \frac{1}{2}. \tag{3.66}$$

From (3.62), (3.65) and (3.66), we see that the autovariance for $\{x_k^2\}$ vanishes.

Derivation of $\mathrm{E}[x_k x_m]$ for logistic map: We consider the case where $k \neq m$. Assume $m = k + n$ for some positive integer n. Since $g(x)$ is an even function, $g^{(n)}(x)$ is also even. Further, since $x g^{(n)}(x)\rho(x)$ is the product of one odd function and two even functions, it is an odd function. Hence, we have, for $n \neq 0$,

$$\mathrm{E}[x_k x_m] = \int_{-\infty}^{\infty} x\; g^{(n)}(x)\rho(x) dx = 0. \tag{3.67}$$

Derivation of $\mathrm{E}[x_k x_{k+\kappa} x_{k+\mu} x_{m+\theta}]$ for logistic map: We consider the case where $\kappa, \mu, \theta \in I^+$ and $0 < \kappa < \mu < \theta$. Thus,

$$\mathrm{E}[x_k x_{k+\kappa} x_{k+\mu} x_{k+\theta}] = \int_{-\infty}^{\infty} x\, g^{(\kappa)}(x)\, g^{(\mu)}(x)\, g^{(\theta)}(x)\, \rho(x) dx. \tag{3.68}$$

Making the substitution $x = \cos\phi$ and using (3.61), the above equation becomes

$$\begin{aligned}
\mathrm{E}[x_k x_{k+\kappa} x_{k+\mu} x_{k+\theta}] &= \frac{1}{\pi}\int_0^{\pi} \cos\phi\, g^{(\kappa)}(\cos\phi)\, g^{(\mu)}(\cos\phi)\, g^{(\theta)}(\cos\phi)\, d\phi \\
&= \frac{-1}{\pi}\int_0^{\pi} \cos\phi\, \cos(2^{\kappa}\phi)\, \cos(2^{\mu}\phi)\, \cos(2^{\theta}\phi)\, d\phi \\
&= \frac{-1}{8\pi}\int_0^{\pi} \big(\cos[(2^{\theta}+2^{\mu}+2^{\kappa}+1)\phi] + \cos[(2^{\theta}+2^{\mu}-2^{\kappa}-1)\phi] \\
&\quad + \cos[(2^{\theta}+2^{\mu}+2^{\kappa}-1)\phi] + \cos[(2^{\theta}+2^{\mu}-2^{\kappa}+1)\phi] \\
&\quad + \cos[(2^{\theta}-2^{\mu}+2^{\kappa}+1)\phi] + \cos[(2^{\theta}-2^{\mu}-2^{\kappa}-1)\phi] \\
&\quad + \cos[(2^{\theta}-2^{\mu}+2^{\kappa}-1)\phi] + \cos[(2^{\theta}-2^{\mu}-2^{\kappa}+1)\phi]\big)\, d\phi \\
&= 0
\end{aligned} \tag{3.69}$$

for $0 < \kappa < \mu < \theta$.

For the chaotic sequence generated by the cubic map $x_{k+1}^2 = 4x_k^3 - 3x_k$, which has the same invariant probability density function as the logistic map, the following properties can be readily shown using the same procedures as in the case of logistic map.

$$\mathrm{E}[x_k] = 0 \tag{3.70}$$

$$\mathrm{cov}[x_k^2, x_m^2] = 0 \tag{3.71}$$

$$\mathrm{E}[x_k x_m] = 0 \tag{3.72}$$

$$\mathrm{E}[x_k x_{k+\kappa} x_{k+\mu} x_{k+\theta}] = 0 \tag{3.73}$$

for $k \neq m$ and $0 < \kappa < \mu < \theta$.

4. Performance Analysis Methods for Non-Coherent Differential Chaos-Shift-Keying Systems

Coherent systems such as the chaos-shift-keying (CSK) system studied in the previous chapter are theoretically better than their non-coherent counterparts in terms of performance in additive white Gaussian noise (AWGN) channels. However, as practical techniques for robust chaos synchronization are not yet available for the required signal-to-noise conditions, the requirement of the receivers to reproduce replicas of chaotic carriers remains a major technical barrier to the practical implementation of coherent systems. Without the need for chaos synchronization, non-coherent systems therefore represent more practical forms of systems, despite their less favorable performance. In this chapter we focus on non-coherent chaos-based communication systems. In general, non-coherent detection can take a variety of forms, but its basic principle is to make use of some distinguishable properties of the transmitted signals, which can be some generic deterministic properties (e.g., return-map based detection [Tse *et al.* (2001)] and maximum-likelihood method [Kisel *et al.* (2001)]), or fabricated by a suitable bit arrangement (e.g., differential CSK (DCSK) [Kolumbán *et al.* (1996)]). In particular, we investigate in depth in this chapter, using the discrete-time baseband model described in Sect. 2.6, the DCSK system under single-user and multi-user environments. Broadly we may classify the implementation of multiple access into two types, namely, time-delay-based [Kolumbán *et al.* (1997c); Kennedy *et al.* (1998)] and permutation-based [Lau *et al.* (2002a)] implementations. In this chapter we present the analysis and performance evaluation of these two types of multiple access as applied to DCSK systems.

4.1 Review of the Differential Chaos-Shift-Keying (DCSK) System

The DCSK system was first proposed by Kolumbán, Vizvari, Schwarz and Abel in 1996 [Kolumbán *et al.* (1996)]. The basic principle in the DCSK system is to make use of a special structural arrangement of each bit to carry digital information. Specifically, each bit duration is divided into two equal time slots and every transmitted symbol is represented by two sets of chaotic signal samples sent in the two slots. The first sample set serves as the reference (reference sample) while the second one carries the data (data sample)

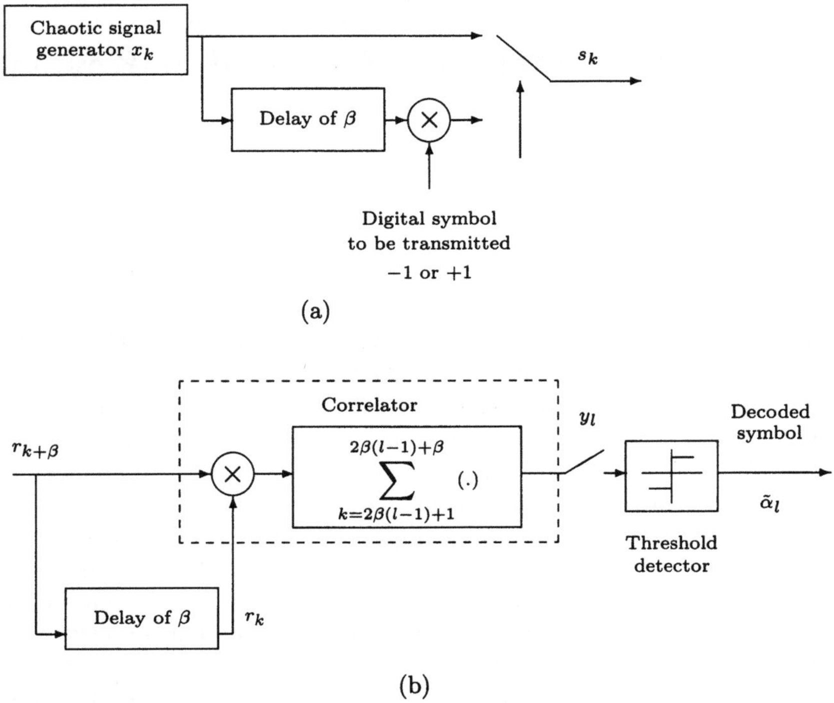

Fig. 4.1. Block diagram of a non-coherent single-user DCSK system. (a) Modulator; (b) demodulator.

[Kolumbán *et al.* (1996); Kolumbán *et al.* (1998a)]. If a "+1" is to be transmitted, the data sample will be identical to the reference sample, and if a "−1" is to be transmitted, an inverted version of the reference sample will be used as the data sample. Figure 4.1 shows the block diagrams of the modulator and demodulator of a noncoherent single-user DCSK system. Let 2β be the spreading factor, defined as the number of chaotic samples sent for each binary symbol, where β is an integer. Also, we denote the lth transmitted symbol by $\alpha_l \in \{-1, +1\}$. During the lth bit duration, i.e., for time $k = 2\beta(l-1)+1, 2\beta(l-1)+2, \ldots, 2\beta l$, the output of the transmitter, s_k, is

$$s_k = \begin{cases} x_k & \text{for } k = 2\beta(l-1)+1, 2\beta(l-1)+2, \ldots, 2\beta(l-1)+\beta \\ \alpha_l x_{k-\beta} & \text{for } k = 2\beta(l-1)+\beta+1, 2\beta(l-1)+\beta+2, \ldots, 2\beta l. \end{cases} \tag{4.1}$$

Suppose the transmitted signal passes through an additive white Gaussian noise (AWGN) channel and arrives at the receiver. At time k, the received signal, r_k, is given by

$$r_k = s_k + \xi_k \tag{4.2}$$

where ξ_k denotes the AWGN with zero mean and variance (power spectral density) $N_0/2$. At the receiving end, the reference sample and the corresponding data sample are correlated. The output of the correlator at the end of the lth symbol duration is given by

$$y_l = \sum_{k=2\beta(l-1)+1}^{2\beta(l-1)+\beta} r_k r_{k+\beta}. \tag{4.3}$$

Depending on whether the output y_l is larger or smaller than zero, a "+1" or "−1" is decoded.

To simplify the notations, we define

$$\begin{aligned}
\boldsymbol{x}_l &= (x_{2\beta(l-1)+1} \;\; x_{2\beta(l-1)+2} \;\; \cdots \;\; x_{2\beta(l-1)+\beta}) && (4.4)\\
\Lambda_m &= (\xi_{\beta m+1} \;\; \xi_{\beta m+2} \;\; \cdots \;\; \xi_{\beta m+\beta}) && (4.5)\\
\boldsymbol{s}_l &= (s_{2\beta(l-1)+1} \;\; s_{2\beta(l-1)+2} \;\; \cdots \;\; s_{2\beta(l-1)+\beta} \;\; s_{2\beta(l-1)+\beta+1} \;\; \cdots \;\; s_{2\beta l}) && (4.6)\\
\boldsymbol{r}_l &= (r_{2\beta(l-1)+1} \;\; r_{2\beta(l-1)+2} \;\; \cdots \;\; r_{2\beta(l-1)+\beta} \;\; r_{2\beta(l-1)+\beta+1} \;\; \cdots \;\; r_{2\beta l}). && (4.7)
\end{aligned}$$

Thus, during the lth symbol duration, the transmitted signal block, $\boldsymbol{s}_l$, can be written as

$$\boldsymbol{s}_l = (\boldsymbol{x}_l \;\; \alpha_l \boldsymbol{x}_l) \tag{4.8}$$

whereas the received signal block, $\boldsymbol{r}_l$, is now represented by

$$\boldsymbol{r}_l = (\boldsymbol{x}_l + \Lambda_{2l-2} \;\; \alpha_l \boldsymbol{x}_l + \Lambda_{2l-1}). \tag{4.9}$$

Further, the correlator output is simplified to

$$y_l = (\boldsymbol{x}_l + \Lambda_{2l-2})(\alpha_l \boldsymbol{x}_l + \Lambda_{2l-1})^T \tag{4.10}$$

where superscript T denotes matrix transposition.

4.2 Multiple Access DCSK System

Before embarking on the issue of multiple access, let us review the basic problem of mutual interference when a number of single-user DCSK systems are operating simultaneously. The situation is depicted by Fig. 4.2. Here, we consider the ith user and denote its lth transmitted symbol by $\alpha_l^{(i)}$. The corresponding signal block is given by

$$\begin{aligned}
\boldsymbol{s}_l^{(i)} &= (s_{2\beta(l-1)+1}^{(i)} \;\; s_{2\beta(l-1)+2}^{(i)} \;\; \cdots \;\; s_{2\beta(l-1)+\beta}^{(i)} \;\; s_{2\beta(l-1)+\beta+1}^{(i)} \;\; \cdots \;\; s_{2\beta l}^{(i)})\\
&= (\boldsymbol{x}_l^{(i)} \;\; \alpha_l^{(i)} \boldsymbol{x}_l^{(i)}) && (4.11)
\end{aligned}$$

where

$$\boldsymbol{x}_l^{(i)} = (x_{2\beta(l-1)+1}^{(i)} \;\; x_{2\beta(l-1)+2}^{(i)} \;\; \cdots \;\; x_{2\beta(l-1)+\beta}^{(i)}) \tag{4.12}$$

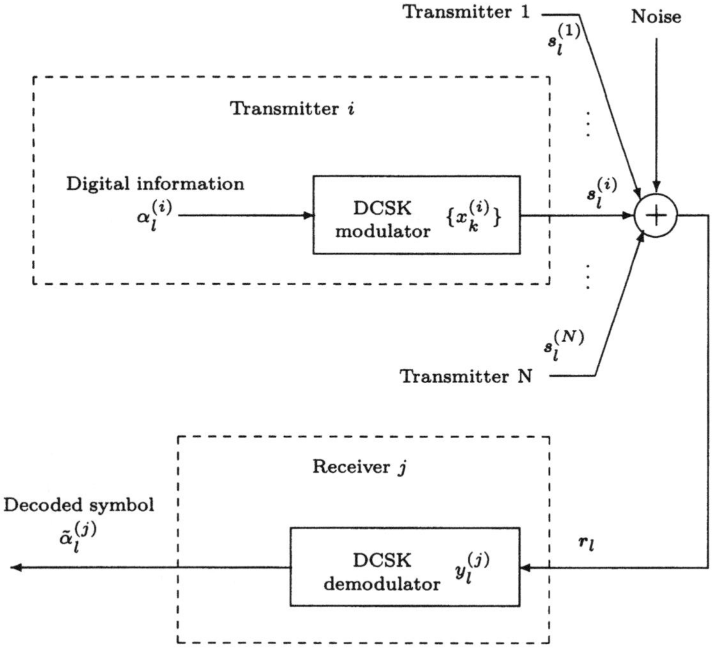

Fig. 4.2. Multiple single-user DCSK systems.

and $\{x_k^{(i)}\}$ represents the chaotic sequence generated by the ith chaos generator. The overall transmitted signal block during the lth bit duration is then

$$\begin{aligned} \boldsymbol{s}_l &= \sum_{i=1}^{N} \boldsymbol{s}_l^{(i)} \\ &= \left(\sum_{i=1}^{N} \boldsymbol{x}_l^{(i)} \quad \sum_{i=1}^{N} \alpha_l^{(i)} \boldsymbol{x}_l^{(i)} \right). \end{aligned} \tag{4.13}$$

At the receiver of the jth user, the lth received signal block is represented by

$$\boldsymbol{r}_l = \left(\sum_{i=1}^{N} \boldsymbol{x}_l^{(i)} + \Lambda_{2l-2} \quad \sum_{i=1}^{N} \alpha_l^{(i)} \boldsymbol{x}_l^{(i)} + \Lambda_{2l-1} \right) \tag{4.14}$$

where Λ_m, as defined in (4.5), denotes the additive noise sample. Similar to the single-user case, the corresponding correlator output, denoted by $y_l^{(j)}$ is found by correlating the first half of the received sample with the second half, i.e.,

$$y_l^{(j)} = \left(\sum_{i=1}^{N} \boldsymbol{x}_l^{(i)} + \Lambda_{2l-2} \right) \left(\sum_{i=1}^{N} \alpha_l^{(i)} \boldsymbol{x}_l^{(i)} + \Lambda_{2l-1} \right)^T$$

$$= \underbrace{\alpha_l^{(j)} \boldsymbol{x}_l^{(j)} (\boldsymbol{x}_l^{(j)})^T}_{\text{required signal}} + \underbrace{\sum_{\substack{i=1 \\ i \neq j}}^{N} \alpha_l^{(i)} \boldsymbol{x}_l^{(i)} (\boldsymbol{x}_l^{(i)})^T}_{\text{strong inter-user interference}}$$

$$+ \underbrace{\sum_{\substack{i=1 \\ i \neq n}}^{N} \sum_{n=1}^{N} \alpha_l^{(n)} \boldsymbol{x}_l^{(i)} (\boldsymbol{x}_l^{(n)})^T}_{\text{weak inter-user interference}}$$

$$+ \underbrace{\sum_{i=1}^{N} \boldsymbol{x}_l^{(i)} \Lambda_{2l-1}^T + \sum_{i=1}^{N} \alpha_l^{(i)} \Lambda_{2l-2} (\boldsymbol{x}_l^{(i)})^T}_{\text{noise}} . \tag{4.15}$$

The first term on the right side of (4.15) is the required signal. The second and third terms represent the strong and weak interference between users, respectively. It can be easily appreciated that since the data sample of each user is identical to the corresponding reference sample or its inversion, strong interference exists in each symbol duration. Thus, a high probability of error for each user is inevitable in this system.

4.3 Time-Delay-Based Multiple Access DCSK System

4.3.1 Frame Structure of the Transmitted Signal

In a multiple access DCSK system, to avoid excessive interference and hence mis-detection, the separation between the reference and data samples should be different for different users. In this section, a time-delay-based multiple access DCSK (TLMA-DCSK) system, in which the separation between the reference and data samples differs for different users, is investigated. The transmission scheme of such a system is illustrated in Fig. 4.3. For all users, each bit duration is divided into 2 slots. For the ith user, $2i$ consecutive slots are collected to form a frame. Hence, the slot duration (half of bit duration) is the same for all users but the frame periods are different for different users. In each frame of the ith user, the first i slots (slots 1 to i) will be used to transmit i sets of reference samples while the remaining i slots (slots $i+1$ to $2i$) are used to transmit i sets of data samples. If a binary symbol "+1" is to be transmitted in slot $i+1$, the samples in slot 1 are repeated in slot $i+1$. If a "-1" is to be sent, an inverted version of the samples in slot 1 will be transmitted in slot $i+1$. Similarly, in slot $i+2$, the same or inverted copy of the samples in slot 2 is sent, and so on. As a result, the reference and data samples of the ith user will be separated by i slots. Therefore, within a frame of length i bit periods (or $2i$ time slots), i bits of information will be sent. This corresponds to an average date rate of 1 bit per bit period. The

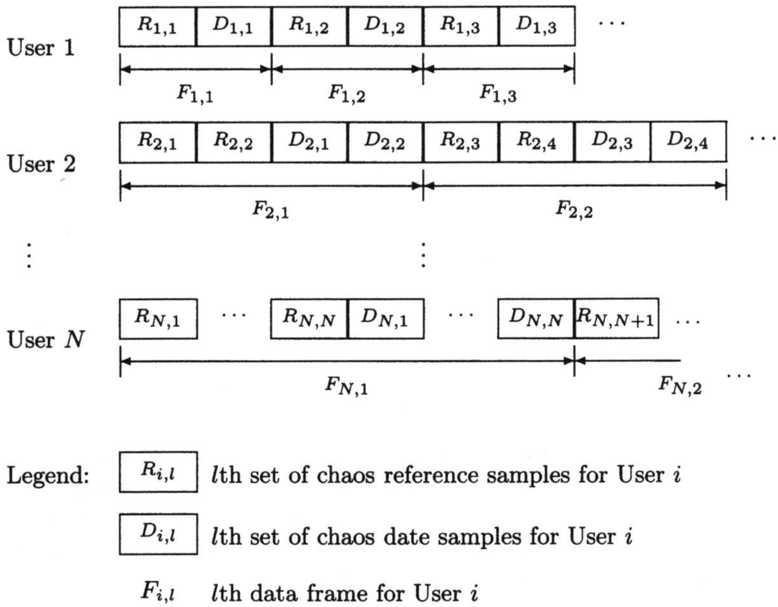

Fig. 4.3. Transmission scheme for the time-delay-based multiple access DCSK (TLMA-DCSK) communication system.

data rates of all users are thus the same. It should be noted that a buffer is required to store the data before transmission.[1]

Suppose there are N users in the system. We define a *superframe* as the minimum number of slots within which all N users transmit an integral number of frames. Within one superframe, we denote the number of frames sent by the ith user by $f_N^{(i)}$. For example, in a 2-user TLMA-DCSK system, a superframe consists of 4 slots, in which User 1 sends two frames and User 2 sends one frame, i.e., $f_2^{(1)} = 2$ and $f_2^{(2)} = 1$. In a 3-user system, each superframe contains 12 slots, and $f_3^{(1)} = 6$, $f_3^{(2)} = 3$ and $f_3^{(3)} = 2$. As all users send a whole number of frames within a superframe, the arrangement of these frames is identical in each superframe. From now on, we restrict our discussion to one superframe which is assumed to contain $2N_s$ slots, where $2N_s = Nf_N^{(N)} = (N-1)f_N^{(N-1)} = \ldots = 2f_N^{(2)} = f_N^{(1)}$.

Figure 4.4 shows a TLMA-DCSK communication system. In the transmitter of the ith user, a chaotic map is used to generate a chaotic sequence $\{x_k^{(i)}\}$ which has a zero mean. The chaotic maps for different users are different in general. Assume that β chaotic samples are sent in each slot (spreading

[1] Note that the frame configuration described here is a particular example from a multitude of possible configurations. For a given number of users, the frame configuration can be optimized so as to minimize the size of buffers required in the transmitters and receivers.

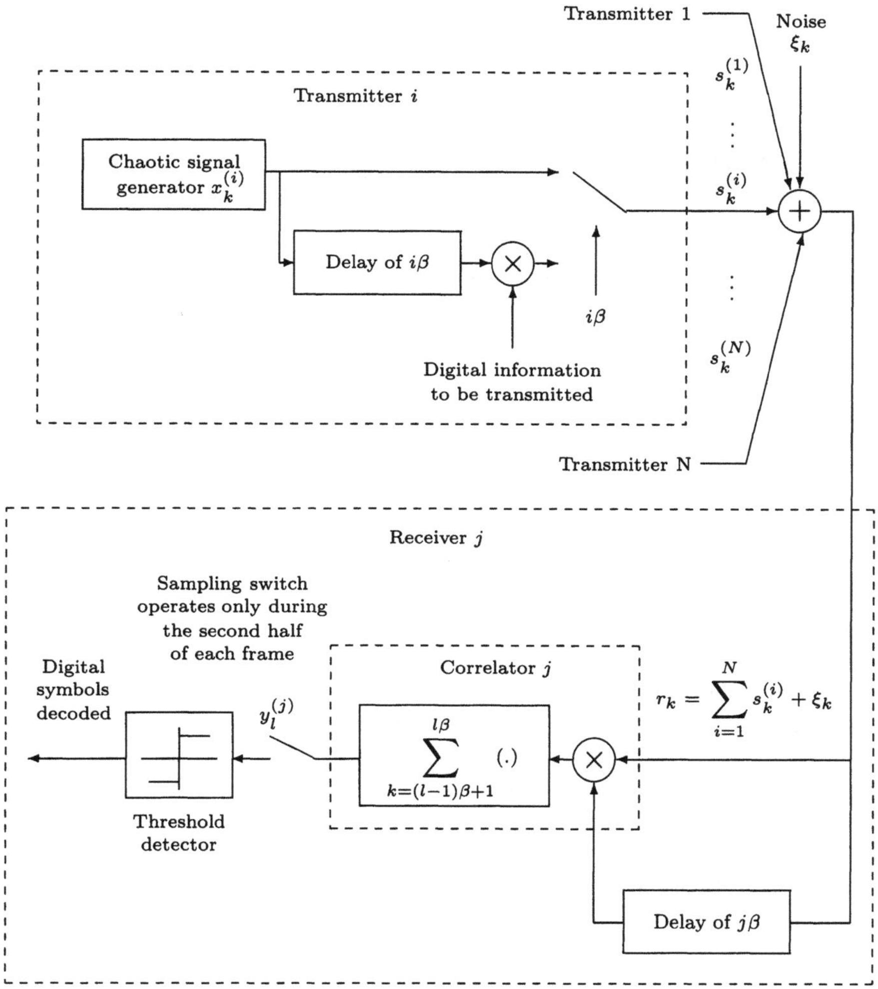

Fig. 4.4. Time-delay-based multiple access DCSK communication system.

factor $= 2\beta$). Consider the transmitted signal of the ith user during the lth time slot, i.e., for time $k = (l-1)\beta + 1, (l-1)\beta + 2, \ldots, l\beta$. Denote the output of the transmitter by $s_k^{(i)}$. If the slot is a reference sample slot, then $s_k^{(i)} = x_k^{(i)}$. If the slot corresponds to a data sample slot sending a binary symbol "+1", we have $s_k^{(i)} = x_{k-i\beta}^{(i)}$. Likewise, if the slot corresponds to a data sample slot sending a binary symbol "−1", we have $s_k^{(i)} = -x_{k-i\beta}^{(i)}$. For notational brevity, for the lth time slot of the ith user, we define $a_l^{(i)}$ as

$$a_l^{(i)} = \begin{cases} +1 & \text{if } y_k^{(i)} = x_{k-\eta}^{(i)} \\ -1 & \text{if } y_k^{(i)} = -x_{k-\eta}^{(i)} \end{cases} \tag{4.16}$$

where $\eta = 0$ for reference samples and $\eta = i\beta$ for data samples. Clearly, the transmitted signal of the ith user during the lth time slot can be represented by

$$s_k^{(i)} = a_l^{(i)} x_{k-\eta}^{(i)}; \quad k = (l-1)\beta + 1, (l-1)\beta + 2, \ldots, l\beta. \tag{4.17}$$

The overall transmitted signal at time k, denoted by s_k, equals

$$s_k = \sum_{i=1}^{N} s_k^{(i)}. \tag{4.18}$$

4.3.2 Receiver Structure

Assuming the channel is AWGN, the received signal at time k, denoted by r_k, is simply given by

$$r_k = s_k + \xi_k \tag{4.19}$$

where ξ_k denotes the AWGN with zero mean and variance (power spectral density) $N_0/2$. For each user, the signal received during a reference sample slot will correlate with the signal at the corresponding data sample slot. Depending on whether the output is larger or smaller than the threshold, a "+1" or "−1" is decoded. Such a correlator-based DCSK receiver is shown in Fig. 4.4. Note that the sampling switch only operates during the second half of each frame and the threshold detector produces a decoded symbol after each slot during that period of time.

The description can be made concise and effective if we consider the whole system as having a *sub-state* at each slot for each particular user. Precisely, denoting the uth sub-state of the system at the lth slot by $\tilde{A}_l^{(u)}$, we write

$$\tilde{A}_l^{(u)} = \left[a_l^{(1)} \; a_l^{(2)} \; \ldots \; a_l^{(u-1)} \; a_l^{(u+1)} \; \ldots \; a_l^{(N)} \right]^T. \tag{4.20}$$

Now consider the jth user and the received signal during the lth time slot. Suppose the slot corresponds to a reference-sample slot for the jth user, i.e., $a_l^{(j)} = +1$. The reference samples in this slot will then correlate with the received samples j slots later, i.e., in the $(l+j)$th slot. The output of the correlator, denoted by $y_l^{(j)}|(\tilde{A}_l^{(j)}, \tilde{A}_{l+j}^{(j)}, a_l^{(j)} = +1, a_{l+j}^{(j)})$, is given by

$$y_l^{(j)}|(\tilde{A}_l^{(j)}, \tilde{A}_{l+j}^{(j)}, a_l^{(j)} = +1, a_{l+j}^{(j)})$$

$$= \sum_{k=(l-1)\beta+1}^{l\beta} \left(\sum_{u=1}^{N} s_k^{(u)} + \xi_k \right) \left(\sum_{v=1}^{N} s_{k+\beta j}^{(v)} + \xi_{k+\beta j} \right)$$

$$\begin{aligned}
&= \sum_{k=(l-1)\beta+1}^{l\beta} \left(\sum_{u=1}^{N} a_l^{(u)} x_{k-\eta_u}^{(u)} + \xi_k \right) \left(\sum_{v=1}^{N} a_{l+j}^{(v)} x_{k+\beta j-\eta_v}^{(v)} + \xi_{k+\beta j} \right) \\
&= \sum_{u=1}^{N}\sum_{v=1}^{N} \sum_{k=(l-1)\beta+1}^{l\beta} a_l^{(u)} a_{l+j}^{(v)} x_{k-\eta_u}^{(u)} x_{k+\beta j-\eta_v}^{(v)} \\
&\quad + \sum_{u=1}^{N} \sum_{k=(l-1)\beta+1}^{l\beta} a_l^{(u)} x_{k-\eta_u}^{(u)} \xi_{k+\beta j} \\
&\quad + \sum_{v=1}^{N} \sum_{k=(l-1)\beta+1}^{l\beta} a_{l+j}^{(v)} x_{k+\beta j-\eta_v}^{(v)} \xi_k + \sum_{k=(l-1)\beta+1}^{l\beta} \xi_k \xi_{k+\beta j}
\end{aligned} \tag{4.21}$$

where $\eta_i = 0$ $(i = 1, 2, \ldots, N)$ for reference samples, and $\eta_i = i\beta$ for data samples. The decoding of the symbol corresponding to this pair of time slots, denoted by $\tilde{\alpha}_l^{(j)}$, can be done according to the following simple rule:

$$\tilde{\alpha}_l^{(j)} = \begin{cases} +1 & \text{if } y_l^{(j)} | (\tilde{A}_l^{(j)}, \tilde{A}_{l+j}^{(j)}, a_l^{(j)} = +1, a_{l+j}^{(j)}) > 0 \\ -1 & \text{if } y_l^{(j)} | (\tilde{A}_l^{(j)}, \tilde{A}_{l+j}^{(j)}, a_l^{(j)} = +1, a_{l+j}^{(j)}) \le 0. \end{cases} \tag{4.22}$$

4.3.3 Derivation of Bit Error Rate

Let the binary symbol transmitted by the jth user in the $(l+j)$th slot be $\alpha_l^{(j)}$, i.e., $a_{l+j}^{(j)} = \alpha_l^{(j)}$. For the jth user, the transmitted signals in the lth slot and $(l+j)$th slot form a reference-data sample pair. Thus, when $u = v = j$, we can simplify (4.21) by putting $x_{k-\eta_u}^{(u)} = x_{k+\beta j-\eta_v}^{(v)} = x_k^{(j)}$. Also, define

$$\tilde{y}_l^{(j)} = y_l^{(j)} | (\tilde{A}_l^{(j)}, \tilde{A}_{l+j}^{(j)}, a_l^{(j)} = +1, \alpha_l^{(j)}) \tag{4.23}$$

$$\begin{aligned}
\tilde{w}_{j,u,v} &= w_{j,u,v} | (\tilde{A}_l^{(j)}, \tilde{A}_{l+j}^{(j)}, a_l^{(j)} = +1, \alpha_l^{(j)}) \\
&= a_l^{(u)} a_{l+j}^{(v)} \sum_{k=(l-1)\beta+1}^{l\beta} x_{k-\eta_u}^{(u)} x_{k+\beta j-\eta_v}^{(v)}
\end{aligned} \tag{4.24}$$

$$\tilde{p}_{j,u} = p_{j,u} | (\tilde{A}_l^{(j)}, a_l^{(j)} = +1) = a_l^{(u)} \sum_{k=(l-1)\beta+1}^{l\beta} x_{k-\eta_u}^{(u)} \xi_{k+\beta j} \tag{4.25}$$

$$\tilde{q}_{j,v} = q_{j,v} | (\tilde{A}_{l+j}^{(j)}, \alpha_l^{(j)}) = a_{l+j}^{(v)} \sum_{k=(l-1)\beta+1}^{l\beta} x_{k+\beta j-\eta_v}^{(v)} \xi_k \tag{4.26}$$

$$\gamma = \sum_{k=(l-1)\beta+1}^{l\beta} \xi_k \xi_{k+\beta j}. \tag{4.27}$$

The correlator output, as given by (4.21), becomes

$$\tilde{y}_l^{(j)} = \underbrace{\tilde{w}_{j,j,j}}_{\text{required signal}} + \underbrace{\sum_{\substack{u=1 \\ (u,v)\neq(j,j)}}^{N}\sum_{v=1}^{N} \tilde{w}_{j,u,v}}_{\text{inter-user interference}} + \underbrace{\sum_{u=1}^{N} \tilde{p}_{j,u} + \sum_{v=1}^{N} \tilde{q}_{j,v} + \gamma}_{\text{noise component}} \tag{4.28}$$

with

$$\tilde{w}_{j,j,j} = \alpha_l^{(j)} \sum_{k=(l-1)\beta+1}^{l\beta} (x_k^{(j)})^2. \tag{4.29}$$

At this point, we make several assumptions which will help simplify the subsequent analysis.

1. Chaotic sequences from different generators are independent of one another.
2. All chaotic sequences have zero mean.
3. All chaotic or noise signals are stationary.

With these assumptions, we can easily show that the mean of $\tilde{y}_l^{(j)}$, $\mathrm{E}[\tilde{y}_l^{(j)}]$, is given by

$$\begin{aligned} \mathrm{E}[\tilde{y}_l^{(j)}] &= \alpha_l^{(j)} \sum_{k=(l-1)\beta+1}^{l\beta} \mathrm{E}[(x_k^{(j)})^2] + \sum_{\substack{u=1 \\ (u,v)\neq(j,j)}}^{N}\sum_{v=1}^{N} \mathrm{E}[\tilde{w}_{j,u,v}] \\ &\quad + \sum_{u=1}^{N} \mathrm{E}[\tilde{p}_{j,u}] + \sum_{v=1}^{N} \mathrm{E}[\tilde{q}_{j,v}] + \mathrm{E}[\gamma] \\ &= \beta\alpha_l^{(j)}\mathrm{E}[(x_k^{(j)})^2] \end{aligned} \tag{4.30}$$

where $\mathrm{E}[\psi]$ denotes the mean of ψ. Also, the variance of $\tilde{y}_l^{(j)}$, denoted by $\mathrm{var}[\tilde{y}_l^{(j)}]$, is

$$\begin{aligned} \mathrm{var}[\tilde{y}_l^{(j)}] &= \mathrm{var}[\tilde{w}_{j,j,j}] + \sum_{\substack{u=1 \\ (u,v)\neq(j,j)}}^{N}\sum_{v=1}^{N} \mathrm{var}[\tilde{w}_{j,u,v}] + \sum_{u=1}^{N} \mathrm{var}[\tilde{p}_{j,u}] + \sum_{v=1}^{N} \mathrm{var}[\tilde{q}_{j,v}] \\ &\quad + \mathrm{var}[\gamma] + \sum_{\substack{C \\ C\neq D}} \sum_{D} \mathrm{cov}[C, D] \end{aligned} \tag{4.31}$$

where $C, D \in \{\tilde{w}_{j,u,v}, \tilde{p}_{j,u}, \tilde{q}_{j,v}, \gamma; u, v = 1, \ldots, N\}$, $\mathrm{var}[\psi]$ denotes the variance of ψ, and $\mathrm{cov}[C, D]$ represents the covariance of C and D which is defined as

$$\mathrm{cov}[C, D] = \mathrm{E}[CD] - \mathrm{E}[C]\mathrm{E}[D]. \tag{4.32}$$

Furthermore, it can be readily shown that the covariance terms in (4.31) vanish, except that when $C = \tilde{w}_{j,u,v}$ and $D = \tilde{w}_{j,\tilde{u},\tilde{v}}$ with $(\tilde{u},\tilde{v}) = (v,u) \neq (u,v)$, the covariance term is given by (see Appendix 4)

$$\begin{aligned}\mathrm{cov}[\tilde{w}_{j,u,v},\tilde{w}_{j,v,u}] &= a_l^{(u)} a_{l+j}^{(u)} a_l^{(v)} a_{l+j}^{(v)} \\ &\times \sum_{k=(l-1)\beta+1}^{l\beta} \sum_{m=(l-1)\beta+1}^{l\beta} \Big(\mathrm{E}[x_{k-\eta_u}^{(u)} x_{m+\beta j-\eta_u}^{(u)}] \\ &\qquad \times \mathrm{E}[x_{m-\eta_v}^{(v)} x_{k+\beta j-\eta_v}^{(v)}]\Big) \end{aligned} \quad (4.33)$$

which is a function of the sub-state of the system. Moreover, the variance terms, as evaluated in Appendix 4, are found to be independent of the sub-state. Thus, (4.31) can be rewritten as

$$\begin{aligned}\mathrm{var}[\tilde{y}_l^{(j)}] &= \mathrm{var}[\tilde{w}_{j,j,j}] + \sum_{\substack{u=1\\(u,v)\neq(j,j)}}^{N} \sum_{v=1}^{N} \mathrm{var}[\tilde{w}_{j,u,v}] + \sum_{u=1}^{N} \mathrm{var}[\tilde{p}_{j,u}] + \sum_{v=1}^{N} \mathrm{var}[\tilde{q}_{j,v}] \\ &+ \mathrm{var}[\gamma] + 2\sum_{\substack{u=1\\u>v}}^{N}\sum_{v=1}^{N} \mathrm{cov}[\tilde{w}_{j,u,v},\tilde{w}_{j,v,u}]. \end{aligned} \quad (4.34)$$

Note that $\tilde{y}_l^{(j)}$ is the sum of a large number of variables. If we assume that its distribution is normal, the probability of an incorrect decision, $\mathrm{BER}_l^{(j)}$, is given by

$$\begin{aligned}&\mathrm{BER}_l^{(j)} \\ &= \sum_{\tilde{A}_l^{(j)} \in U^{(N-1)}} \sum_{\tilde{A}_{l+j}^{(j)} \in U^{(N-1)}} \Big[\mathrm{Prob}(\tilde{y}_l^{(j)} \leq 0 \mid \alpha_l^{(j)} = +1) \\ &\qquad \times \mathrm{Prob}(\tilde{A}_l^{(j)}, \tilde{A}_{l+j}^{(j)}) \times \mathrm{Prob}(\alpha_l^{(j)} = +1) \\ &\qquad + \mathrm{Prob}(\tilde{y}_l^{(j)} > 0 \mid \alpha_l^{(j)} = -1) \\ &\qquad \times \mathrm{Prob}(\tilde{A}_l^{(j)}, \tilde{A}_{l+j}^{(j)}) \times \mathrm{Prob}(\alpha_l^{(j)} = -1)\Big] \\ &= \frac{1}{2} \sum_{\tilde{A}_l^{(j)} \in U^{(N-1)}} \sum_{\tilde{A}_{l+j}^{(j)} \in U^{(N-1)}} \left[\frac{1}{2}\mathrm{erfc}\left(\frac{\mathrm{E}\left[\tilde{y}_l^{(j)} | \alpha_l^{(j)} = +1\right]}{\sqrt{2\mathrm{var}\left[\tilde{y}_l^{(j)} | \alpha_l^{(j)} = +1\right]}}\right)\right. \\ &\qquad \left. + \frac{1}{2}\mathrm{erfc}\left(\frac{-\mathrm{E}\left[\tilde{y}_l^{(j)} | \alpha_l^{(j)} = -1\right]}{\sqrt{2\mathrm{var}\left[\tilde{y}_l^{(j)} | \alpha_l^{(j)} = -1\right]}}\right)\right] \\ &\qquad \times \mathrm{Prob}(\tilde{A}_l^{(j)}, \tilde{A}_{l+j}^{(j)}) \end{aligned} \quad (4.35)$$

where $U \in \{-1, +1\}$ and erfc(.) is the complementary error function defined as

$$\mathrm{erfc}(\psi) \equiv \frac{2}{\sqrt{\pi}} \int_{\psi}^{\infty} e^{-\lambda^2} d\lambda. \tag{4.36}$$

Define $S^{(j)}$ as the set of slot numbers corresponding to the reference slots of the jth user over one superframe. The average BER for the jth user, $\mathrm{BER}^{(j)}$, is then found by summing all the possible values of l over one superframe, i.e.,

$$\begin{aligned}\mathrm{BER}^{(j)} &= \frac{1}{2N_s} \sum_{l \in S^{(j)}} \sum_{\tilde{A}_l^{(j)} \in U^{(N-1)}} \sum_{\tilde{A}_{l+j}^{(j)} \in U^{(N-1)}} \\ &\quad \left[\frac{1}{2}\mathrm{erfc}\left(\frac{\mathrm{E}\left[\tilde{y}_l^{(j)} | \alpha_l^{(j)} = +1\right]}{\sqrt{2\mathrm{var}\left[\tilde{y}_l^{(j)} | \alpha_l^{(j)} = +1\right]}} \right) \right. \\ &\quad \left. + \frac{1}{2}\mathrm{erfc}\left(\frac{-\mathrm{E}\left[\tilde{y}_l^{(j)} | \alpha_l^{(j)} = -1\right]}{\sqrt{2\mathrm{var}\left[\tilde{y}_l^{(j)} | \alpha_l^{(j)} = -1\right]}} \right) \right] \\ &\quad \times \mathrm{Prob}(\tilde{A}_l^{(j)}, \tilde{A}_{l+j}^{(j)}). \end{aligned} \tag{4.37}$$

Furthermore, define $B(\tilde{A}_l^{(j)}, \tilde{A}_{l+j}^{(j)})$ as the average probability of occurrence of $(\tilde{A}_l^{(j)}, \tilde{A}_{l+j}^{(j)})$ over all reference slots, i.e.,

$$B(\tilde{A}_l^{(j)}, \tilde{A}_{l+j}^{(j)}) = \frac{1}{N_s} \sum_{l \in S^{(j)}} \mathrm{Prob}(\tilde{A}_l^{(j)}, \tilde{A}_{l+j}^{(j)}). \tag{4.38}$$

Thus, (4.37) can be rewritten as

$$\begin{aligned}\mathrm{BER}^{(j)} &= \frac{1}{2} \sum_{\tilde{A}_l^{(j)} \in U^{(N-1)}} \sum_{\tilde{A}_{l+j}^{(j)} \in U^{(N-1)}} \left[\frac{1}{2}\mathrm{erfc}\left(\frac{\mathrm{E}\left[\tilde{y}_l^{(j)} | \alpha_l^{(j)} = +1\right]}{\sqrt{2\mathrm{var}\left[\tilde{y}_l^{(j)} | \alpha_l^{(j)} = +1\right]}} \right) \right. \\ &\quad \left. + \frac{1}{2}\mathrm{erfc}\left(\frac{-\mathrm{E}\left[\tilde{y}_l^{(j)} | \alpha_l^{(j)} = -1\right]}{\sqrt{2\mathrm{var}\left[\tilde{y}_l^{(j)} | \alpha_l^{(j)} = -1\right]}} \right) \right] \\ &\quad \times B(\tilde{A}_l^{(j)}, \tilde{A}_{l+j}^{(j)}). \end{aligned} \tag{4.39}$$

Each of the sub-states $\tilde{A}_l^{(j)}$ and $\tilde{A}_{l+j}^{(j)}$ has $2^{(N-1)}$ possible combinations. Hence, $(\tilde{A}_l^{(j)}, \tilde{A}_{l+j}^{(j)})$ has a total of $2^{(2N-2)}$ possible combinations. The average probability of occurrence of each of these combinations, $B(\tilde{A}_l^{(j)}, \tilde{A}_{l+j}^{(j)})$, can be derived for the jth user by studying the structure of the superframe. For example, in a 3-user system, there are 16 combinations, and the corresponding probabilities for User 1, denoted by $\mathbf{B}_3^{(1)}$, can be shown equal to

$$\mathbf{B}_3^{(1)} = \frac{1}{96}[\ 35\ \ 15\ \ 7\ \ 3\ \ 5\ \ 5\ \ 1\ \ 1\ \ 7\ \ 3\ \ 7\ \ 3\ \ 1\ \ 1\ \ 1\ \ 1]^T. \tag{4.40}$$

Similarly, the corresponding probabilities for User 2 ($\mathbf{B}_3^{(2)}$) and User 3 ($\mathbf{B}_3^{(3)}$) can be shown equal to

$$\mathbf{B}_3^{(2)} = \frac{1}{96}[\ 37\ \ 11\ \ 5\ \ 3\ \ 11\ \ 1\ \ 3\ \ 1\ \ 5\ \ 3\ \ 5\ \ 3\ \ 3\ \ 1\ \ 3\ \ 1\ ]^T \tag{4.41}$$

$$\mathbf{B}_3^{(3)} = \frac{1}{48}[\ 13\ \ 5\ \ 8\ \ 4\ \ 5\ \ 1\ \ 4\ \ 0\ \ 5\ \ 1\ \ 0\ \ 0\ \ 1\ \ 1\ \ 0\ \ 0\ ]^T. \tag{4.42}$$

The first element in each array corresponds to $B([+1,+1],[+1,+1])$, and the second one corresponds to $B([+1,+1],[+1,-1])$, ..., and the last element corresponds to $B([-1,-1],[-1,-1])$. Note that due to the asymmetrical structure of the superframe, the probabilities are different for different users. Moreover, elements in each array all add up to 1.

4.3.3.1 Mixed Analysis-Simulation Technique. Without making specific assumptions about the kinds of chaotic maps used, we can evaluate the BERs by a mixed analysis-simulation (MAS) approach. Specifically, from (4.30) and (4.34), the mean, variance and covariance terms can be obtained by simulations. Hence, (4.30), (4.34) and (4.39) provide a straightforward mixed analysis-simulation technique to derive the BERs of the system.

4.3.3.2 Analytical BERs. If further assumptions are made, closed-form analytical BERs can be obtained. In particular, in order to ensure that all users are treated equally, we assume that all users transmit with equal average power, i.e.,

$$P_s = P_s^{(1)} = P_s^{(2)} = \cdots = P_s^{(N)} \tag{4.43}$$

where

$$P_s^{(i)} = \mathrm{E}[(x_k^{(i)})^2]. \tag{4.44}$$

The equation in (4.30) becomes

$$\mathrm{E}[\tilde{y}_l^{(j)}] = \beta\alpha_l^{(j)} P_s. \tag{4.45}$$

Special Case I: Chaotic sequences with particular statistical properties

Assume that the chaotic sequence generated by the ith user, $\{x_k^{(i)}\}$, satisfies the following conditions.

$$\begin{aligned}
\mathrm{cov}[(x_k^{(i)})^2, (x_m^{(i)})^2] &= \mathrm{E}[(x_k^{(i)})^2 (x_m^{(i)})^2] \\
&\quad - \mathrm{E}[(x_k^{(i)})^2]\mathrm{E}[(x_m^{(i)})^2] = 0 \quad k \neq m \qquad (4.46) \\
\mathrm{E}[x_k^{(i)} x_m^{(i)}] &= 0 \quad k \neq m \qquad (4.47) \\
\mathrm{E}[x_k^{(i)} x_{k+\kappa}^{(i)} x_{k+\mu}^{(i)} x_{k+\theta}^{(i)}] &= 0 \quad \forall\, \kappa, \mu, \theta \in I^+ \text{ and } 0 < \kappa < \mu < \theta \qquad (4.48)
\end{aligned}$$

The first two assumptions make the autovariance of $\{(x_k^{(i)})^2\}$ and the autocorrelation of $\{x_k^{(i)}\}$ vanish. The third one essentially implies that any four unequal numbers taken from the sequence $\{x_k^{(i)}\}$ must be uncorrelated. Thus, we have

$$\begin{aligned}
\mathrm{var}[\tilde{y}_l^{(j)}] &= \beta \mathrm{var}[(x_k^{(j)})^2] + \sum_{\substack{u=1\\(u,v)\neq(j,j)}}^{N} \sum_{v=1}^{N} \beta P_s^2 + \sum_{u=1}^{N} \beta P_s N_0/2 + \sum_{v=1}^{N} \beta P_s N_0/2 \\
&\quad + \beta N_0^2/4 + 2 \sum_{\substack{u=1\\u>v}}^{N} \sum_{v=1}^{N} (0) \\
&= \beta \mathrm{var}[(x_k^{(j)})^2] + \beta(N^2-1)P_s^2 + \beta N P_s N_0 + \beta N_0^2/4. \qquad (4.49)
\end{aligned}$$

Furthermore, under such circumstances, $\mathrm{E}[\tilde{y}_l^{(j)}|\alpha_l^{(j)} = +1] = -\mathrm{E}[\tilde{y}_l^{(j)}|\alpha_l^{(j)} = -1]$ and $\mathrm{var}[\tilde{y}_l^{(j)}]$ is independent of $\alpha_l^{(j)}$. Moreover, $\mathrm{E}[\tilde{y}_l^{(j)}]$ and $\mathrm{var}[\tilde{y}_l^{(j)}]$ are now independent of $\tilde{A}_l^{(j)}$ and $\tilde{A}_{l+j}^{(j)}$. Therefore, (4.39) becomes

$$\begin{aligned}
\mathrm{BER}_{\text{DCSK-I}}^{(j)} &= \frac{1}{2}\left[\frac{1}{2}\mathrm{erfc}\left(\frac{\mathrm{E}\left[\tilde{y}_l^{(j)}|\alpha_l^{(j)} = +1\right]}{\sqrt{2\mathrm{var}\left[\tilde{y}_l^{(j)}|\alpha_l^{(j)} = +1\right]}}\right)\right. \\
&\quad \left. + \frac{1}{2}\mathrm{erfc}\left(\frac{-\mathrm{E}\left[\tilde{y}_l^{(j)}|\alpha_l^{(j)} = -1\right]}{\sqrt{2\mathrm{var}\left[\tilde{y}_l^{(j)}|\alpha_l^{(j)} = -1\right]}}\right)\right] \\
&\quad \times \sum_{\tilde{A}_l^{(j)} \in U^{(N-1)}} \sum_{\tilde{A}_{l+j}^{(j)} \in U^{(N-1)}} B(\tilde{A}_l^{(j)}, \tilde{A}_{l+j}^{(j)}) \\
&= \frac{1}{2}\mathrm{erfc}\left(\frac{\mathrm{E}\left[\tilde{y}_l^{(j)}|\alpha_l^{(j)} = +1\right]}{\sqrt{2\mathrm{var}\left[\tilde{y}_l^{(j)}|\alpha_l^{(j)} = +1\right]}}\right) \\
&= \frac{1}{2}\mathrm{erfc}\left(\left[\frac{2\mathrm{var}[(x_k^{(j)})^2]}{\beta P_s^2} + \frac{2(N^2-1)}{\beta}\right.\right. \\
&\quad \left.\left. + \frac{2NN_0}{\beta P_s} + \frac{N_0^2}{2\beta P_s^2}\right]^{-\frac{1}{2}}\right). \qquad (4.50)
\end{aligned}$$

The second equality holds because $\sum_{\tilde{A}_l^{(j)} \in U^{(N-1)}} \sum_{\tilde{A}_{l+j}^{(j)} \in U^{(N-1)}} B(\tilde{A}_l^{(j)}, \tilde{A}_{l+j}^{(j)})$ is equal to one. For notational convenience, we define

$$\Psi^{(j)} = \frac{\mathrm{var}[(x_k^{(j)})^2]}{P_s^2} = \frac{\mathrm{var}[(x_k^{(j)})^2]}{\mathrm{E}^2[(x_k^{(j)})^2]} \tag{4.51}$$

which depends only on the type of the chaotic sequence used and not on the average power of the sequence (see Sect. 3.2.3.2 for details). The BER for the jth user can be written as

$$\mathrm{BER}_{\text{DCSK-I}}^{(j)} = \frac{1}{2}\mathrm{erfc}\left(\left[\frac{2\Psi^{(j)}}{\beta} + \frac{2(N^2-1)}{\beta} + \frac{2NN_0}{\beta P_s} + \frac{N_0^2}{2\beta P_s^2}\right]^{-\frac{1}{2}}\right) \tag{4.52}$$

$$= \frac{1}{2}\mathrm{erfc}\left(\left[\frac{2\Psi^{(j)}}{\beta} + \frac{2(N^2-1)}{\beta} + 4N\left(\frac{E_b}{N_0}\right)^{-1} + 2\beta\left(\frac{E_b}{N_0}\right)^{-2}\right]^{-\frac{1}{2}}\right) \tag{4.53}$$

where E_b denotes the average bit energy, i.e.,

$$E_b = 2\beta P_s. \tag{4.54}$$

We further define

$$\theta_{\text{DCSK-I}} = \underbrace{\frac{2\Psi^{(j)}}{\beta}}_{\text{variation of } (x_k^{(j)})^2} + \underbrace{\frac{2(N^2-1)}{\beta}}_{\text{inter-user interference}} + \underbrace{\frac{2NN_0}{\beta P_s} + \frac{N_0^2}{2\beta P_s^2}}_{\text{noise component}}$$

$$= \frac{2\Psi^{(j)}}{\beta} + \frac{2(N^2-1)}{\beta} + 4N\left(\frac{E_b}{N_0}\right)^{-1} + 2\beta\left(\frac{E_b}{N_0}\right)^{-2}. \tag{4.55}$$

Clearly, the performance of the TLMA-DCSK system is affected by

1. spreading factor 2β;
2. variance of $\{(x_k^{(j)})^2\}$ for a given P_s;
3. number of users N; and
4. noise power.

Thus, for a fixed E_b/N_0, we may improve the BER for the jth user by making one or more of the following adjustments.

1. Change the spreading factor 2β until the optimal BER is obtained [Abel *et al.* (2000), Kolumbán (2000b)].
2. Minimize the variance of $(x_k^{(j)})^2$ for a fixed P_s. In the optimal case where $\mathrm{var}[(x_k^{(j)})^2] = 0$, binary sequences will be produced.
3. Reduce the number of users N.

Special Case II: All users use the same chaotic map to generate sequences with particular statistical properties

Again we assume that the chaotic sequences satisfy the conditions given in (4.46) to (4.48). In addition, we assume that all users employ the same map, but each uses a different initial condition. Thus, the statistical properties of $\{x_k^{(i)}\}$ will be identical for all users. We may write the BER for each user as

$$\begin{aligned} \text{BER}_{\text{DCSK-II}} &= \frac{1}{2}\text{erfc}\left(\left[\frac{2\Psi}{\beta}+\frac{2(N^2-1)}{\beta}+\frac{2NN_0}{\beta P_s}+\frac{N_0^2}{2\beta P_s^2}\right]^{-\frac{1}{2}}\right) \quad (4.56)\\ &= \frac{1}{2}\text{erfc}\left(\left[\frac{2\Psi}{\beta}+\frac{2(N^2-1)}{\beta}+4N\left(\frac{E_b}{N_0}\right)^{-1}\right.\right.\\ &\qquad\left.\left. +2\beta\left(\frac{E_b}{N_0}\right)^{-2}\right]^{-\frac{1}{2}}\right) \quad (4.57)\end{aligned}$$

where

$$\Psi=\frac{\text{var}[(x_k^{(1)})^2]}{P_s^2}=\frac{\text{var}[(x_k^{(2)})^2]}{P_s^2}=\cdots=\frac{\text{var}[(x_k^{(N)})^2]}{P_s^2}. \quad (4.58)$$

As in the previous case, for a fixed E_b/N_0, we can improve the BER of the system by making one or more of the following adjustments.

1. Change the spreading factor 2β until the optimal BER is obtained.
2. Minimize the variance of $(x_k^{(j)})^2$ for a fixed P_s.
3. Reduce the number of users N.

4.3.4 Simulations and Evaluation

The following chaotic maps are used in our simulations.

1. Logistic map
$$x_{k+1}=1-2x_k^2 \quad (4.59)$$
2. Cubic map
$$x_{k+1}=4x_k^3-3x_k \quad (4.60)$$
3. Hénon map
$$\begin{cases} x_{k+1} = 1+y_k-1.4x_k^2\\ y_{k+1} = 0.3x_k\end{cases} \quad (4.61)$$
4. Bernoulli-shift map
$$x_{k+1}=\begin{cases}1.2x_k+1 & \text{when } x_k<0\\ 1.2x_k-1 & \text{when } x_k>0\end{cases} \quad (4.62)$$

Table 4.1. Statistical properties of the normalized chaotic sequences

	Logistic map	Cubic map	Hénon map	Bernoulli-shift map
$\mathrm{E}[x_k]$	0	0	0	0
$\mathrm{E}[x_k^2]$	1	1	1	1
$\mathrm{E}[x_k^3]$	0	0	−0.4950	0
$\mathrm{var}[x_k]$	1	1	1	1
$\mathrm{var}[x_k^2]$	0.5	0.5	1.1216	0.9874

In our simulations, the chaotic sequences generated by (4.59) to (4.62) are normalized before they are transmitted. This ensures that the mean value of the transmitted signal is zero and the mean-squared is one, i.e., $\mathrm{E}[x_k] = 0$ and $\mathrm{E}[x_k^2] = 1$. The statistical properties of the normalized sequences are shown in Table 4.1. It can be shown (see Appendix 3B) that the chaotic sequences generated from the logistic map and the cubic map satisfy the conditions stated in Sect. 4.3.3.2, i.e.,

$$\begin{aligned}
\mathrm{cov}[x_k^2, x_m^2] &= \mathrm{E}[x_k^2 x_m^2] - \mathrm{E}[x_k^2]\mathrm{E}[x_m^2] = 0 \quad k \neq m && (4.63)\\
\mathrm{E}[x_k x_m] &= 0 \quad k \neq m && (4.64)\\
\mathrm{E}[x_k x_{k+\kappa} x_{k+\mu} x_{k+\theta}] &= 0 \quad \forall\, \kappa, \mu, \theta \in I^+ \text{ and } 0 < \kappa < \mu < \theta. && (4.65)
\end{aligned}$$

4.3.4.1 Users Using Distinct Chaotic Maps. We now study a 3-user TLMA-DCSK system where each user uses a distinct chaotic map:

User 1: Logistic map
User 2: Cubic map
User 3: Hénon map

The spreading factor is 200 ($\beta = 100$). The BERs predicted using the mixed analysis-simulation (MAS) approach described in Sect. 4.3.3.1 and from the analytical expressions derived in Sect. 4.3.3.2 are shown in Figs. 4.5 to 4.7. For comparison and verification, we also present the BERs calculated from brute-force simulations which give the true system performance.[2]

It can be observed that the users using the logistic map and the cubic map (i.e., Users 1 and 2) have similar BER performance, and the user using the Hénon map (i.e., User 3) has a slightly better BER performance.

[2] System performance can be simulated using the empirical system model without making simplifying assumptions of the chaotic maps or the forms of the various mean, variance and co-variance terms. Simulations performed in this way can be considered as "brute-force" simulations, and they provide true system performance for comparison and verification purposes. See also p. 52.

4.3.4.2 Same Chaotic Map Used by All Users. Next we study a 3-user TLMA-DCSK system in which all chaotic sequences are generated from the cubic map (4.60) and a spreading factor of 200 is used ($\beta = 100$). The BERs found from analytical expressions, MAS approach and brute-force simulations are shown in Fig. 4.8.

It is interesting to compare the simulated BERs for different choices of chaotic maps. Figure 4.9 shows the comparative performance when the logistic map, cubic map and Bernoulli-shift map are used. It can be observed that the Bernoulli-shift map has a higher BER due to the larger var$[x_k^2]$.

Finally, we study the effect of the choice of spreading factor on the BER. Specifically, in Fig. 4.10, we plot the analytical BERs as given by (4.57) versus the spreading factor (2β) for different values of E_b/N_0 for a 3-user system and a 5-user system when the logistic map is used. From Fig. 4.10, we see that the BER improves initially as the spreading factor increases, but deteriorates as the spreading factor continues to increase beyond a certain optimal value.

Before closing this section, it is worth reiterating that the foregoing analysis has assumed the time slots of different users being synchronized. However, this assumption is not necessarily valid in practice. Nonetheless, as long as all users use chaotic sequences that satisfy the conditions stated in Sect. 4.3.3.2, the level of mutual interference between users should not be affected even if the time slots among users are not synchronized.

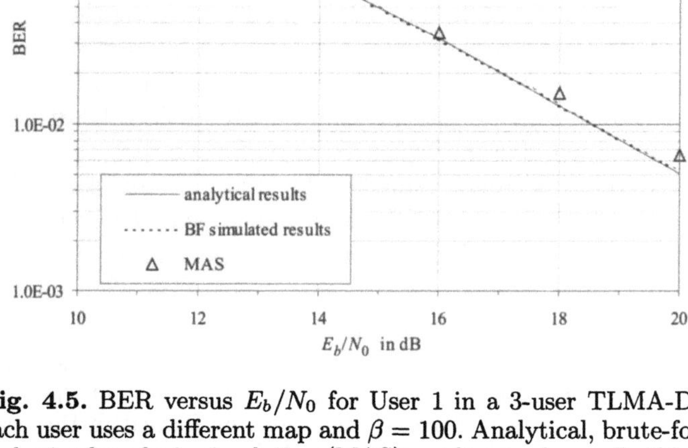

Fig. 4.5. BER versus E_b/N_0 for User 1 in a 3-user TLMA-DCSK system where each user uses a different map and $\beta = 100$. Analytical, brute-force (BF) simulation and mixed analysis-simulation (MAS) results are shown. In this system, User 1 uses the logistic map, User 2 uses the cubic map and User 3 uses the Hénon map.

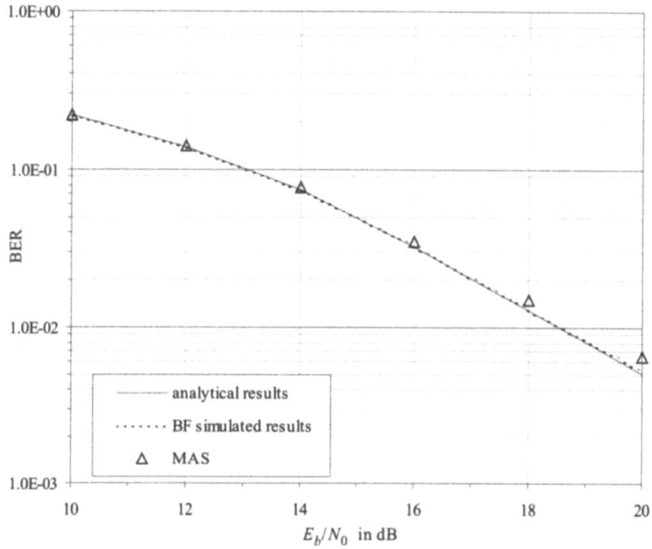

Fig. 4.6. BER versus E_b/N_0 for User 2 in a 3-user TLMA-DCSK system where each user uses a different map and $\beta = 100$. Analytical, brute-force (BF) simulation and mixed analysis-simulation (MAS) results are shown. In this system, User 1 uses the logistic map, User 2 uses the cubic map and User 3 uses the Hénon map.

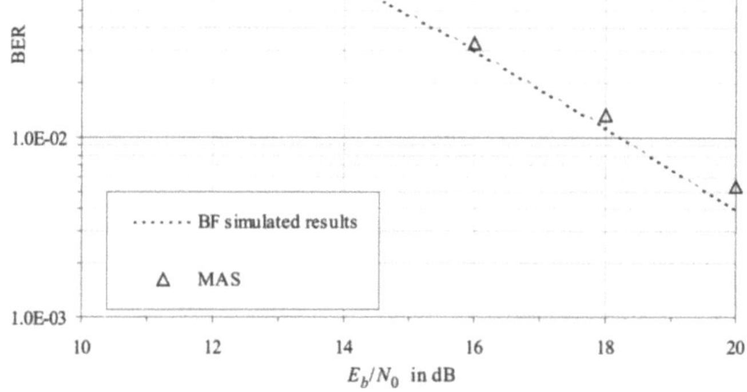

Fig. 4.7. BER versus E_b/N_0 for User 3 in a 3-user TLMA-DCSK system where each user uses a different map and $\beta = 100$. Brute-force (BF) simulation and mixed analysis-simulation (MAS) results are shown. In this system, User 1 uses the logistic map, User 2 uses the cubic map and User 3 uses the Hénon map.

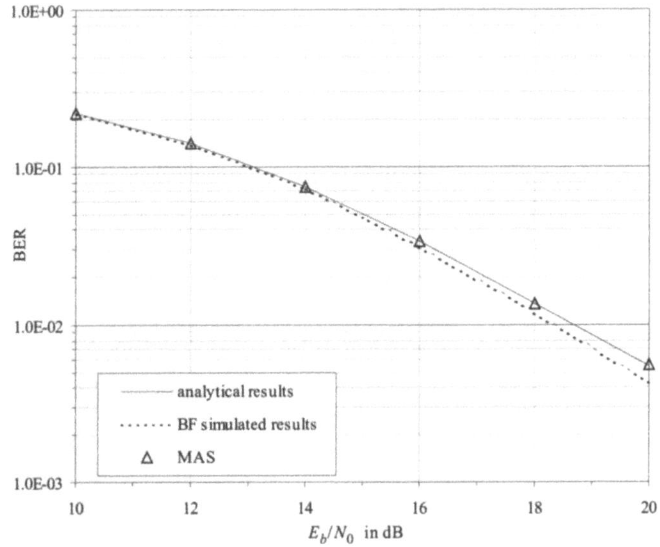

Fig. 4.8. BER versus E_b/N_0 in a 3-user TLMA-DCSK system where all users use the cubic map and $\beta = 100$. Analytical, brute-force (BF) simulation and mixed analysis-simulation (MAS) results are shown.

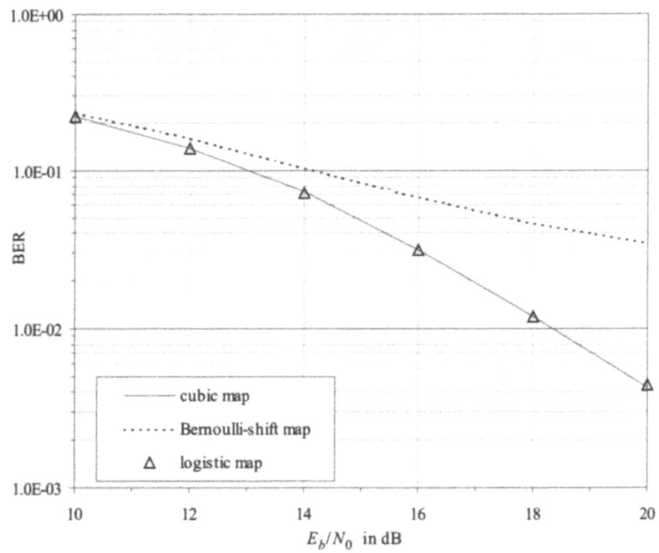

Fig. 4.9. BER versus E_b/N_0 by brute-force simulation in a 3-user TLMA-DCSK system where all users use the same chaotic map and $\beta = 100$. Three cases are simulated, corresponding to the use of the logistic map, cubic map and Bernoulli-shift map as the chaos generator.

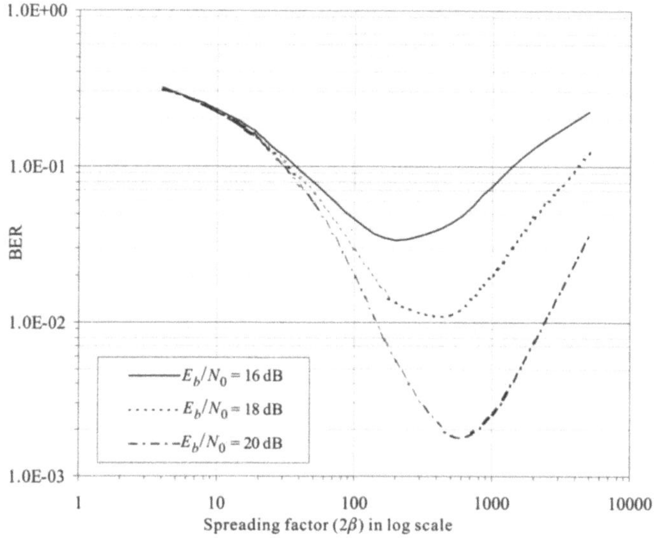

(a)

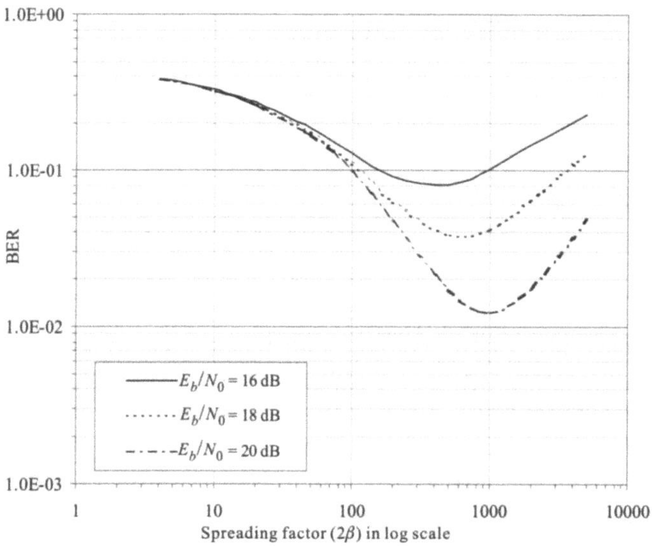

(b)

Fig. 4.10. Analytical BER versus spreading factor (2β) under different E_b/N_0 in a TLMA-DCSK system where all users use the logistic map as the chaos generator. (a) 3-user system; (b) 5-user system.

4.4 Permutation-Based Multiple Access DCSK System

4.4.1 System Description

An alternative approach to the implementation of multiple access systems is to apply permutation to the chaotic sequences, with the same objective of minimizing interference among users. Figure 4.11 shows the block diagram of a permutation-based multiple access DCSK (PMA-DCSK) system. Using the same notation as in Sect. 4.2, we denote the signal block generated by the DCSK modulator of the ith user by $\boldsymbol{s}_l^{(i)}$. The main feature of the PMA-DCSK scheme is that each signal block of the ith user undergoes a transformation F_i before it is transmitted. This transformation essentially destroys the similarity between the reference and data samples, leading to possible enhancement of the data security of the system. Specifically, the transmitted signal block, $\boldsymbol{u}_l^{(i)}$, is given by

$$\begin{aligned} \boldsymbol{u}_l^{(i)} &= (u_{2\beta(l-1)+1}^{(i)} \; u_{2\beta(l-1)+2}^{(i)} \; \cdots \; u_{2\beta(l-1)+\beta}^{(i)} \; u_{2\beta(l-1)+\beta+1}^{(i)} \; \cdots \; u_{2\beta l}^{(i)}) \\ &= F_i(\boldsymbol{s}_l^{(i)}). \end{aligned} \quad (4.66)$$

The overall transmitted signal block of the whole system during the lth bit duration, denoted by $\boldsymbol{u}_l$, is therefore

$$\boldsymbol{u}_l = \sum_{i=1}^{N} \boldsymbol{u}_l^{(i)} = \sum_{i=1}^{N} F_i(\boldsymbol{s}_l^{(i)}). \quad (4.67)$$

At the receiver of the jth user, the lth received signal block is $\boldsymbol{v}_l = \boldsymbol{u}_l + (\Lambda_{2l-2} \; \Lambda_{2l-1})$, which will undergo an inverse transformation $G_j = F_j^{-1}$, resulting in an output signal block $\boldsymbol{r}_l^{(j)}$ which is given by

$$\begin{aligned} \boldsymbol{r}_l^{(j)} &= (r_{2\beta(l-1)+1}^{(j)} \; r_{2\beta(l-1)+2}^{(j)} \; \cdots \; r_{2\beta(l-1)+\beta}^{(j)} \; r_{2\beta(l-1)+\beta+1}^{(j)} \; \cdots \; r_{2\beta l}^{(j)}) \\ &= G_j(\boldsymbol{v}_l). \end{aligned} \quad (4.68)$$

This output signal block $\boldsymbol{r}_l^{(j)}$ is then processed by the usual DCSK demodulator. Clearly, in this PMA-DCSK system, additional transformation and inverse transformation are needed in the transmitter and receiver, respectively. The main objective of the additional transformations is to eliminate the strong inter-user interference term in (4.15).

The choice of the transformation is crucial to the performance of the system. Here, the transformation process is a simple permutation operation. Specifically, for the ith user, the transformation and inverse transformation are given by

$$F_i(\boldsymbol{s}_l^{(i)}) = \boldsymbol{s}_l^{(i)} \boldsymbol{P}_{2\beta}^{(i)} \quad (4.69)$$

$$G_i(\boldsymbol{v}_l) = F_i^{-1}(\boldsymbol{v}_l) = \boldsymbol{v}_l(\boldsymbol{P}_{2\beta}^{(i)})^{-1} = \boldsymbol{v}_l(\boldsymbol{P}_{2\beta}^{(i)})^T \quad (4.70)$$

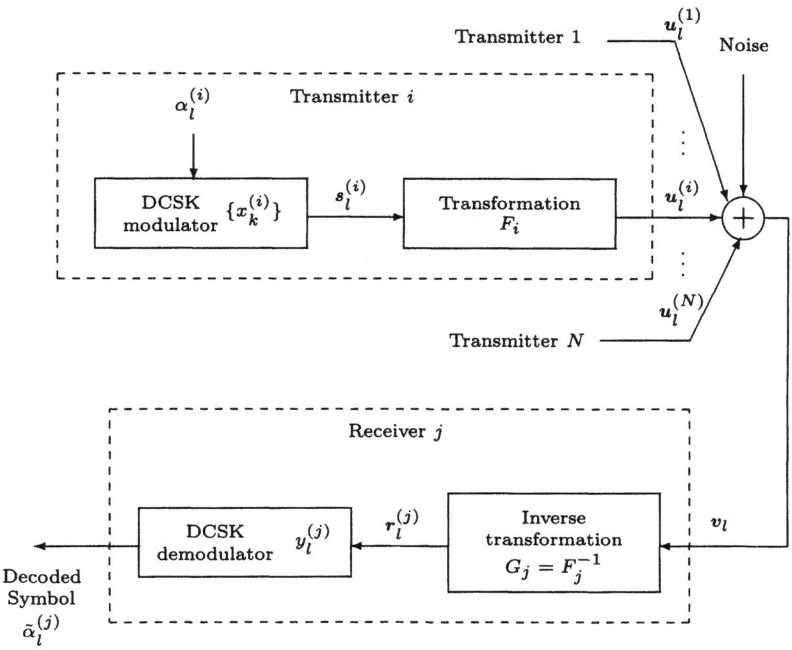

Fig. 4.11. Permutation-based Multiple Access DCSK communication system.

where $P_{2\beta}^{(i)}$ is a $2\beta \times 2\beta$ *permutation matrix*.[3] Also, since the inverse of a permutation matrix is equal to its transpose, the output from the inverse transformation is

$$
\begin{aligned}
r_l^{(j)} &= G_j(v_l) \\
&= (u_l + (\Lambda_{2l-2} \quad \Lambda_{2l-1}))(P_{2\beta}^{(j)})^T \\
&= u_l(P_{2\beta}^{(j)})^T + (\Lambda_{2l-2} \quad \Lambda_{2l-1})(P_{2\beta}^{(j)})^T \\
&= \sum_{i=1}^{N} s_l^{(i)} P_{2\beta}^{(i)} (P_{2\beta}^{(j)})^T + (\Lambda_{2l-2} \quad \Lambda_{2l-1})(P_{2\beta}^{(j)})^T \\
&= s_l^{(j)} + \sum_{\substack{i=1 \\ i \neq j}}^{N} s_l^{(i)} P_{2\beta}^{(i)} (P_{2\beta}^{(j)})^T + (\Lambda_{2l-2} \quad \Lambda_{2l-1})(P_{2\beta}^{(j)})^T \\
&= \underbrace{s_l^{(j)}}_{\text{required DCSK signal of user } j} + \underbrace{\sum_{\substack{i=1 \\ i \neq j}}^{N} s_l^{(i)} P_{2\beta}^{(i,j)}}_{\text{interference from other users}}
\end{aligned}
$$

[3] The permutation matrix is defined as a square matrix whose elements are either "0" or "1", with each row and column containing exactly one "1" [Stewart (1998)].

$$+ \underbrace{(\Lambda_{2l-2} \quad \Lambda_{2l-1})(\boldsymbol{P}_{2\beta}^{(j)})^T}_{\text{noise}} \tag{4.71}$$

where

$$\boldsymbol{P}_{2\beta}^{(i,j)} = \boldsymbol{P}_{2\beta}^{(i)}(\boldsymbol{P}_{2\beta}^{(j)})^T. \tag{4.72}$$

Note that $\boldsymbol{P}_{2\beta}^{(i,j)}$ is the product of two permutation matrices, and is therefore also a permutation matrix. To ensure a low inter-user interference, the $(\lambda + \beta)$th $(\lambda = 1, 2, \ldots, \beta)$ element in $\boldsymbol{s}_l^{(i)}\boldsymbol{P}_{2\beta}^{(i,j)}$ $(i \neq j)$ should not equal to the λth element or its inversion. In other words, in $\boldsymbol{P}_{2\beta}^{(i,j)}$, the non-zero elements in the λth $(\lambda = 1, 2, \ldots, \beta)$ and $(\lambda + \beta)$th rows should not differ by β columns. For convenience, we denote by $p_{a,b}^{(i,j)}$ the element in the ath row and bth column in the matrix $\boldsymbol{P}_{2\beta}^{(i,j)}$. If the non-zero elements in the λth and $(\lambda+\beta)$th rows are $p_{\lambda,\mu}^{(i,j)}$ and $p_{\lambda+\beta,\nu}^{(i,j)}$, respectively, then $|\mu - \nu| \neq \beta$. Clearly there are many ways to choose the permutation matrices such that the aforementioned condition is satisfied. In the following, we describe a simple technique to accomplish this.

We define $\boldsymbol{R}_\beta$ as a random permutation matrix of size $\beta \times \beta$, and $\boldsymbol{A}_\beta$ as a "shifting" matrix of size $\beta \times \beta$, i.e.,

$$\boldsymbol{A}_\beta = \begin{pmatrix} 0 & 1 & 0 & \cdots & 0 \\ 0 & 0 & 1 & \ddots & \vdots \\ \vdots & \ddots & \ddots & \ddots & 0 \\ 0 & \ddots & \ddots & \ddots & 1 \\ 1 & 0 & \cdots & 0 & 0 \end{pmatrix}. \tag{4.73}$$

Note that the consequence of multiplying $\boldsymbol{A}_\beta$ to an array $(x_1 \ x_2 \ \ldots \ x_\beta)$ k times is equivalent to shifting the elements in the array to the right k times with the overflown elements being re-inserted from the left. In other words,

$$(x_1 \ x_2 \ \ldots \ x_\beta)\boldsymbol{A}_\beta^k = (x_{\beta-k+1} \ x_{\beta-k+2} \ \ldots \ x_\beta \ x_1 \ x_2 \ \ldots \ x_{\beta-k}). \tag{4.74}$$

Now, the permutation matrix for the ith user can be chosen as

$$\boldsymbol{P}_{2\beta}^{(i)} = \begin{pmatrix} \boldsymbol{I}_\beta & \boldsymbol{0}_\beta \\ \boldsymbol{0}_\beta & \boldsymbol{R}_\beta\boldsymbol{A}_\beta^i \end{pmatrix} \tag{4.75}$$

where $\boldsymbol{I}_\beta$ and $\boldsymbol{0}_\beta$ are, respectively, the identity matrix and zero matrix of size $\beta \times \beta$. Also, (4.72) can be re-written as

$$\boldsymbol{P}_{2\beta}^{(i,j)} = \begin{pmatrix} \boldsymbol{I}_\beta & \boldsymbol{0}_\beta \\ \boldsymbol{0}_\beta & \boldsymbol{R}_\beta\boldsymbol{A}_\beta^{i-j}\boldsymbol{R}_\beta^T \end{pmatrix}. \tag{4.76}$$

It can be readily shown that the diagonal elements of $\boldsymbol{R}_\beta\boldsymbol{A}_\beta^{i-j}\boldsymbol{R}_\beta^T$ are all zeros when $i \neq j$. Thus, the non-zero elements in the λth $(\lambda = 1, 2, \ldots, \beta)$

and $(\lambda+\beta)$th rows in $\boldsymbol{P}_{2\beta}^{(i,j)}$ will not differ by β columns, and a low inter-user interference is achieved. Substituting (4.11), (4.75) and (4.76) into (4.71), we obtain

$$\begin{aligned} \boldsymbol{r}_l^{(j)} &= (\boldsymbol{x}_l^{(j)} \quad \alpha_l^{(j)}\boldsymbol{x}_l^{(j)}) + \left(\sum_{\substack{i=1\\i\neq j}}^{N} \boldsymbol{x}_l^{(i)} \quad \sum_{\substack{i=1\\i\neq j}}^{N} \alpha_l^{(i)}\boldsymbol{x}_l^{(i)}\boldsymbol{R}_\beta \boldsymbol{A}_\beta^{i-j}\boldsymbol{R}_\beta^T \right) \\ &\quad + \left(\Lambda_{2l-2} \quad \Lambda_{2l-1}\boldsymbol{A}_\beta^{-j}\boldsymbol{R}_\beta^T\right) \\ &= (\Gamma \;\; \Theta) \end{aligned} \tag{4.77}$$

where

$$\Gamma = \boldsymbol{x}_l^{(j)} + \sum_{\substack{i=1\\i\neq j}}^{N} \boldsymbol{x}_l^{(i)} + \Lambda_{2l-2} \tag{4.78}$$

$$\Theta = \alpha_l^{(j)}\boldsymbol{x}_l^{(j)} + \sum_{\substack{i=1\\i\neq j}}^{N} \alpha_l^{(i)}\boldsymbol{x}_l^{(i)}\boldsymbol{R}_\beta\boldsymbol{A}_\beta^{i-j}\boldsymbol{R}_\beta^T + \Lambda_{2l-1}\boldsymbol{A}_\beta^{-j}\boldsymbol{R}_\beta^T. \tag{4.79}$$

Suppose $\boldsymbol{r}_l^{(j)}$ is sent to the conventional DCSK demodulator of the jth user. The output of the correlator at the end of the lth bit duration, denoted by $y_l^{(j)}$, can be computed from

$$y_l^{(j)} = \Gamma\Theta^T. \tag{4.80}$$

The decoding rule is simply given by

$$\tilde{\alpha}_l^{(j)} = \begin{cases} +1 & \text{if } y_l^{(j)} > 0 \\ -1 & \text{if } y_l^{(j)} \leq 0. \end{cases} \tag{4.81}$$

where $\tilde{\alpha}_l^{(j)}$ is the lth decoded symbol for the jth user.

4.4.2 Derivation of Bit Error Rate

Without loss of generality, we consider the probability of error for the first symbol (i.e., $l = 1$) transmitted by the jth user. For brevity, the subscripts of the variables $y_l^{(j)}, \boldsymbol{x}_l^{(i)}$ and $\alpha_l^{(i)}$ are omitted. The input to the threshold detector at the end of the first symbol duration is given by

$$y^{(j)} = \alpha^{(j)}\boldsymbol{x}^{(j)}(\boldsymbol{x}^{(j)})^T + \sum_{\substack{i=1\\i\neq j}}^{N} \alpha^{(i)}\boldsymbol{x}^{(j)}\boldsymbol{R}_\beta\boldsymbol{A}_\beta^{j-i}\boldsymbol{R}_\beta^T(\boldsymbol{x}^{(i)})^T$$

$$
\begin{aligned}
& + \alpha^{(j)} \sum_{\substack{i=1 \\ i \neq j}}^{N} \boldsymbol{x}^{(i)} (\boldsymbol{x}^{(j)})^T + \sum_{\substack{n=1 \\ n \neq j}}^{N} \sum_{\substack{i=1 \\ i \neq j}}^{N} \alpha^{(i)} \boldsymbol{x}^{(n)} \boldsymbol{R}_\beta \boldsymbol{A}_\beta^{j-i} \boldsymbol{R}_\beta^T (\boldsymbol{x}^{(i)})^T \\
& + \sum_{i=1}^{N} \boldsymbol{x}^{(i)} \boldsymbol{R}_\beta \boldsymbol{A}_\beta^{j} \Lambda_1^T + \sum_{i=1}^{N} \alpha^{(i)} \Lambda_0 \boldsymbol{R}_\beta \boldsymbol{A}_\beta^{j-i} \boldsymbol{R}_\beta^T (\boldsymbol{x}^{(i)})^T + \Lambda_0 \boldsymbol{R}_\beta \boldsymbol{A}_\beta^{j} \Lambda_1^T \\
= \; & \alpha^{(j)} \boldsymbol{x}^{(j)} (\boldsymbol{x}^{(j)})^T + \sum_{\substack{i=1 \\ i \neq j}}^{N} \alpha^{(i)} \boldsymbol{x}^{(j)} (\boldsymbol{z}^{(i,j)})^T \\
& + \alpha^{(j)} \sum_{\substack{i=1 \\ i \neq j}}^{N} \boldsymbol{x}^{(i)} (\boldsymbol{x}^{(j)})^T + \sum_{\substack{n=1 \\ n \neq j}}^{N} \sum_{\substack{i=1 \\ i \neq j}}^{N} \alpha^{(i)} \boldsymbol{x}^{(n)} (\boldsymbol{z}^{(i,j)})^T \\
& + \sum_{i=1}^{N} \boldsymbol{x}^{(i)} (\boldsymbol{\Phi}^{(j)})^T + \sum_{i=1}^{N} \alpha^{(i)} \Lambda_0 (\boldsymbol{z}^{(i,j)})^T + \Lambda_0 (\boldsymbol{\Phi}^{(j)})^T \qquad (4.82)
\end{aligned}
$$

where

$$
\begin{aligned}
\boldsymbol{z}^{(i,j)} &= (z_1^{(i,j)} \; z_2^{(i,j)} \; \cdots \; z_\beta^{(i,j)}) = \boldsymbol{x}^{(i)} \boldsymbol{R}_\beta \boldsymbol{A}_\beta^{i-j} \boldsymbol{R}_\beta^T \qquad (4.83) \\
\boldsymbol{\Phi}^{(j)} &= (\phi_1^{(j)} \; \phi_2^{(j)} \; \cdots \; \phi_\beta^{(j)}) = \Lambda_1 \boldsymbol{A}_\beta^{-j} \boldsymbol{R}_\beta^T. \qquad (4.84)
\end{aligned}
$$

Note that the elements in $\boldsymbol{z}^{(i,j)}$ and $\boldsymbol{\Phi}^{(j)}$ are permutations of the elements in $\boldsymbol{x}^{(i)}$ and Λ_1, respectively.

Suppose "+1" is transmitted for the jth user, i.e., $\alpha^{(j)} = +1$. Then, (4.82) becomes

$$
\begin{aligned}
y^{(j)} | (\alpha^{(j)} = +1) = \; & \underbrace{U^{(j,j)}}_{\text{required signal}} \\
& + \underbrace{\sum_{\substack{i=1 \\ i \neq j}}^{N} \alpha^{(i)} V^{(j,i,j)} + \sum_{\substack{i=1 \\ i \neq j}}^{N} U^{(i,j)} + \sum_{\substack{n=1 \\ n \neq j}}^{N} \sum_{\substack{i=1 \\ i \neq j}}^{N} \alpha^{(i)} V^{(n,i,j)}}_{\text{inter-user interference}} \\
& + \underbrace{\sum_{i=1}^{N} W^{(i,j)} + X^{(j,j)} + \sum_{\substack{i=1 \\ i \neq j}}^{N} \alpha^{(i)} X^{(i,j)} + Y^{(j)}}_{\text{noise}} \qquad (4.85)
\end{aligned}
$$

where

$$
U^{(i,j)} = \boldsymbol{x}^{(i)} (\boldsymbol{x}^{(j)})^T = \sum_{k=1}^{\beta} x_k^{(i)} x_k^{(j)} \qquad (4.86)
$$

$$
V^{(n,i,j)} = \boldsymbol{x}^{(n)} (\boldsymbol{z}^{(i,j)})^T = \sum_{k=1}^{\beta} x_k^{(n)} z_k^{(i,j)} \qquad (4.87)
$$

$$W^{(i,j)} = \boldsymbol{x}^{(i)}(\boldsymbol{\Phi}^{(j)})^T = \sum_{k=1}^{\beta} x_k^{(i)}\phi_k^{(j)} \tag{4.88}$$

$$X^{(i,j)} = \Lambda_0(\boldsymbol{z}^{(i,j)})^T = \sum_{k=1}^{\beta} \xi_k z_k^{(i,j)} \tag{4.89}$$

$$Y^{(j)} = \Lambda_0(\boldsymbol{\Phi}^{(j)})^T = \sum_{k=1}^{\beta} \xi_k \phi_k^{(j)}. \tag{4.90}$$

Note that the input to the detector consists of three components, namely, the required signal, the inter-user interference and noise. The mean value of $y^{(j)}|(\alpha^{(j)} = +1)$ is given by

$$\begin{aligned}
\mathrm{E}[y^{(j)}|(\alpha^{(j)} = +1)] &= \mathrm{E}[U^{(j,j)}] + \sum_{\substack{i=1\\ i\neq j}}^{N} \alpha^{(i)}\mathrm{E}[V^{(j,i,j)}] + \sum_{\substack{i=1\\ i\neq j}}^{N} \mathrm{E}[U^{(i,j)}] \\
&\quad + \sum_{\substack{n=1\\ n\neq j}}^{N}\sum_{\substack{i=1\\ i\neq j}}^{N} \alpha^{(i)}\mathrm{E}[V^{(n,i,j)}] + \sum_{i=1}^{N}\mathrm{E}[W^{(i,j)}] \\
&\quad + \mathrm{E}[X^{(j,j)}] + \sum_{\substack{i=1\\ i\neq j}}^{N} \alpha^{(i)}\mathrm{E}[X^{(i,j)}] + \mathrm{E}[Y^{(j)}] \\
&= \mathrm{E}[U^{(j,j)}] + \sum_{\substack{i=1\\ i\neq j}}^{N} \alpha^{(i)}\mathrm{E}[V^{(j,i,j)}] + \sum_{\substack{i=1\\ i\neq j}}^{N} \mathrm{E}[U^{(i,j)}] \\
&\quad + \sum_{\substack{n=1\\ n\neq j}}^{N}\sum_{\substack{i=1\\ i\neq j}}^{N} \alpha^{(i)}\mathrm{E}[V^{(n,i,j)}].
\end{aligned} \tag{4.91}$$

The last equality holds because $\mathrm{E}[\xi_k]$ and $\mathrm{E}[\phi_k^{(j)}]$ are both zero. The variance of $y^{(j)}|(\alpha^{(j)} = +1)$ can be found from [Ross (1993)]

$$\begin{aligned}
\mathrm{var}[y^{(j)}|(\alpha^{(j)} = +1)] &= \sum_{i=1}^{N}\mathrm{var}[U^{(i,j)}] + \sum_{n=1}^{N}\sum_{\substack{i=1\\ i\neq j}}^{N}\mathrm{var}[V^{(n,i,j)}] \\
&\quad + \sum_{i=1}^{N}\mathrm{var}[W^{(i,j)}] + \sum_{i=1}^{N}\mathrm{var}[X^{(i,j)}] \\
&\quad + \mathrm{var}[Y^{(j)}] + \sum_{\substack{C\\ C\neq D}}\sum_{D}\mathrm{cov}[C,D]
\end{aligned} \tag{4.92}$$

where $C, D \in \{U^{(i,j)}\ (i = 1,\ldots,N); \alpha^{(i)}V^{(n,i,j)}\ (n = 1,\ldots,N; i = 1,\ldots,j-1, j+1,\ldots,N); W^{(i,j)}\ (i = 1,\ldots,N); X^{(j,j)}; \alpha^{(i)}X^{(i,j)}(i = 1,\ldots,j-1,j+1,\ldots,N); Y^{(j)}\}$.

To illustrate the key analysis procedure, we consider the case where the cubic map

$$x_{k+1} = g(x_k) = 4x_k^3 - 3x_k \tag{4.93}$$

is used by all users to generate the chaotic sequences. Moreover, each user uses a different initial condition. For this map, the mean value of each chaotic sequence is zero (see Appendix 3B), and the following can be easily verified.

$$\begin{aligned} \mathrm{E}[U^{(i,j)}] &= \begin{cases} \beta \mathrm{E}[x_k^2] & \text{if } i=j \\ 0 & \text{otherwise} \end{cases} & (4.94) \\ \mathrm{E}[V^{(n,i,j)}] &= 0 \quad n=1,\ldots,N; i=1,\ldots,j-1,j+1,\ldots,N & (4.95) \end{aligned}$$

Also, the variances of the terms in (4.86) to (4.90) are given by

$$\begin{aligned} \mathrm{var}[U^{(i,j)}] &= \begin{cases} \beta \mathrm{var}[x_k^2] & \text{if } i=j \\ \beta \mathrm{E}^2[x_k^2] & \text{otherwise} \end{cases} & (4.96) \\ \mathrm{var}[V^{(n,i,j)}] &= \begin{cases} \beta \mathrm{E}^2[x_k^2] & \text{for } i \neq j, i \neq n \\ \frac{\beta^2}{\beta-1}\mathrm{E}^2[x_k^2] \approx \beta \mathrm{E}^2[x_k^2] & \text{for } i \neq j, i = n \end{cases} & (4.97) \\ \mathrm{var}[W^{(i,j)}] &= \beta N_0 \mathrm{E}[x_k^2]/2 & (4.98) \\ \mathrm{var}[X^{(i,j)}] &= \beta N_0 \mathrm{E}[x_k^2]/2 & (4.99) \\ \mathrm{var}[Y^{(j)}] &= \beta N_0^2/4 & (4.100) \end{aligned}$$

and all covariance terms in (4.92) are zero. Thus, (4.91) and (4.92) can be simplified to

$$\begin{aligned} \mathrm{E}[y^{(j)}|(\alpha^{(j)}=+1)] &= \beta \mathrm{E}[x_k^2] & (4.101) \\ \mathrm{var}[y^{(j)}|(\alpha^{(j)}=+1)] &\approx \beta \mathrm{var}[x_k^2] + (N-1)\beta \mathrm{E}^2[x_k^2] + (N^2-N)\beta \mathrm{E}^2[x_k^2] \\ &\quad + N\beta N_0 \mathrm{E}[x_k^2]/2 + N\beta N_0 \mathrm{E}[x_k^2]/2 + \beta N_0^2/4 \\ &= \beta \mathrm{var}[x_k^2] + (N^2-1)\beta \mathrm{E}^2[x_k^2] + N\beta N_0 \mathrm{E}[x_k^2] \\ &\quad + \beta N_0^2/4. & (4.102) \end{aligned}$$

Likewise, we can find the mean and variance of $y^{(j)}$ given a "-1" has been sent. Some straightforward algebraic derivations lead to

$$\begin{aligned} \mathrm{E}[y^{(j)}|(\alpha^{(j)}=-1)] &= -\mathrm{E}[y^{(j)}|(\alpha^{(j)}=+1)] & (4.103) \\ \mathrm{var}[y^{(j)}|(\alpha^{(j)}=-1)] &= \mathrm{var}[y^{(j)}|(\alpha^{(j)}=+1)]. & (4.104) \end{aligned}$$

Since $y^{(j)}|(\alpha^{(j)}=+1)$ and $y^{(j)}|(\alpha^{(j)}=-1)$ are the sum of a large number of random variables, we may assume that their distributions are approximately normal. Clearly, this assumption holds better for larger spreading factors [Proakis and Salehi (1994)]. In detecting the symbol, an error occurs if $y^{(j)} \leq 0|(\alpha^{(j)}=+1)$ or $y^{(j)} > 0|(\alpha^{(j)}=-1)$. The BER for the jth user can thus be computed from

$$
\begin{aligned}
\mathrm{BER}^{(j)} &= \frac{1}{2}\mathrm{Prob}(y^{(j)} \le 0|(\alpha^{(j)} = +1)) + \frac{1}{2}\mathrm{Prob}(y^{(j)} > 0|(\alpha^{(j)} = -1)) \\
&= \frac{1}{4}\mathrm{erfc}\left(\frac{\mathrm{E}[y^{(j)}|(\alpha^{(j)} = +1)]}{\sqrt{2\mathrm{var}[y^{(j)}|(\alpha^{(j)} = +1)]}}\right) \\
&\quad + \frac{1}{4}\mathrm{erfc}\left(\frac{-\mathrm{E}[y^{(j)}|(\alpha^{(j)} = -1)]}{\sqrt{2\mathrm{var}[y^{(j)}|(\alpha^{(j)} = -1)]}}\right) \\
&= \frac{1}{2}\mathrm{erfc}\left(\frac{\mathrm{E}[y^{(j)}|(\alpha^{(j)} = +1)]}{\sqrt{2\mathrm{var}[y^{(j)}|(\alpha^{(j)} = +1)]}}\right) \\
&= \frac{1}{2}\mathrm{erfc}\left(\frac{\beta\mathrm{E}[x_k^2]}{\sqrt{2\left(\begin{array}{c}\beta\mathrm{var}[x_k^2] + (N^2-1)\beta\mathrm{E}^2[x_k^2] + N\beta N_0\mathrm{E}[x_k^2] \\ +\,\beta N_0^2/4\end{array}\right)}}\right) \\
&= \frac{1}{2}\mathrm{erfc}\left(\left[\frac{2\Psi}{\beta} + \frac{2(N^2-1)}{\beta} + \frac{2NN_0}{\beta P_s} + \frac{N_0^2}{2\beta P_s^2}\right]^{-\frac{1}{2}}\right) \qquad (4.105) \\
&= \frac{1}{2}\mathrm{erfc}\left(\left[\frac{2\Psi}{\beta} + \frac{2(N^2-1)}{\beta} + 4N\left(\frac{E_b}{N_0}\right)^{-1}\right.\right. \\
&\qquad\qquad \left.\left. + 2\beta\left(\frac{E_b}{N_0}\right)^{-2}\right]^{-\frac{1}{2}}\right) \qquad (4.106)
\end{aligned}
$$

where erfc(.), E_b and Ψ have been defined earlier in (4.36), (4.54) and (4.58), respectively. Note that the BER expression obtained in (4.106) is independent of j, meaning that all users are having the same BER performance.

4.4.3 Simulations and Evaluation

In this section, the performance of the permutation-based multiple access DCSK (PMA-DCSK) digital communication system is studied by computer simulations. We will use the cubic map as the chaos generator. Also, we assume that all users generate the chaotic sequences using this map, and each uses a different initial condition. For this map, it can be shown that

$$
\begin{aligned}
\mathrm{E}[x_k^2] &= 0.5 \qquad (4.107) \\
\mathrm{var}[x_k^2] &= 0.125. \qquad (4.108)
\end{aligned}
$$

The analytical BER can thus be obtained by substituting (4.107) and (4.108) into (4.106). In particular, the BER performance of the PMA-DCSK communication system will be investigated under variation of the following parameters:

– average-bit-energy-to-noise-power-spectral-density ratio (E_b/N_0)

– number of users (N)
– spreading factor (2β).

Also of interest is the performance comparison with the time-delay-based multiple access DCSK (TLMA-DCSK) system studied earlier in Sect. 4.3.

The relevant BERs for the DCSK systems are shown in Figs. 4.12 to 4.14. From Fig. 4.12, we clearly see that the analytical and simulated BERs are in very good agreement for a spreading factor of 200, where the assumption of normal distribution of the conditional correlator output holds well. Further observations are summarized as follows.

1. The PMA-DCSK and TLMA-DCSK systems have very similar performance. Comparing (4.57) and (4.106), their analytical BER performances are the same.
2. BERs increase (degrade) as the number of users increases for a given E_b/N_0. This is apparently due to the increasing inter-user interference.

In concluding this chapter we reiterate that the MA-DCSK schemes have been studied with the assumption that the time slots are synchronized among all participating users. Nonetheless, we expect that the interference between users will not vary too much even when the time slots are not synchronized. In other words, the actual BER performance should not deviate substantially from that found in this chapter.

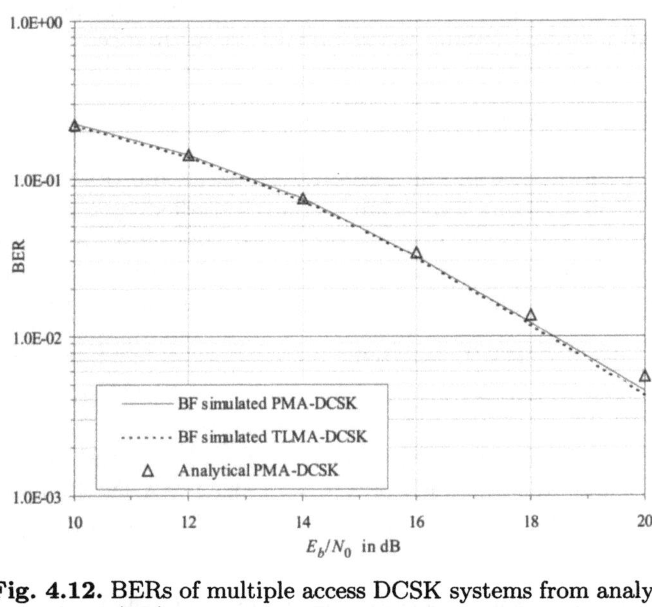

Fig. 4.12. BERs of multiple access DCSK systems from analytical expressions and brute-force (BF) simulations. Spreading factor $2\beta = 200$ and $N = 3$.

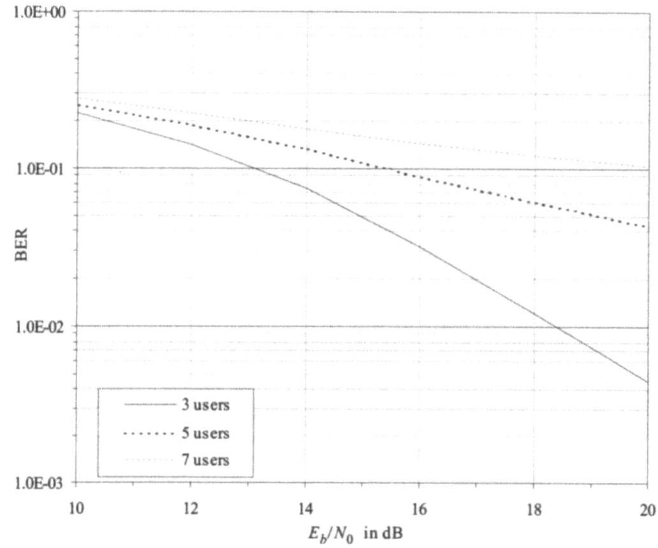

Fig. 4.13. BERs versus E_b/N_0 for the PMA-DCSK system from brute-force simulations. Number of users $N = 3$, 5 and 7. Spreading factor $2\beta = 200$.

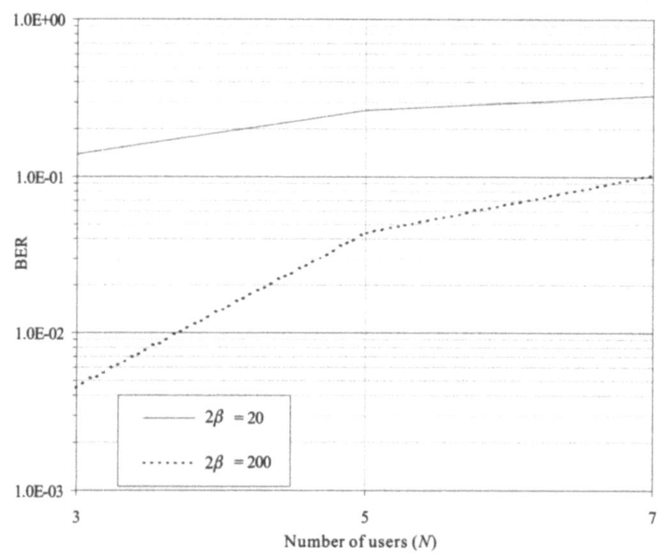

Fig. 4.14. BERs versus number of users N for the PMA-DCSK system from brute-force simulations. Spreading factor $2\beta = 20$ and 200. $E_b/N_0 = 20$ dB.

Appendix 4: Derivation of variances and covariances relevant to the analysis of time-delay-based multiple access DCSK system

All symbols are as defined in Sect. 4.3.

1. The derivation of the variance of $\tilde{w}_{j,j,j}$ is as follows.

$$\begin{aligned}
\mathrm{var}[\tilde{w}_{j,j,j}] &= \mathrm{var}\left[\alpha_l^{(j)} \sum_{k=(l-1)\beta+1}^{l\beta} (x_k^{(j)})^2\right] \\
&= \sum_{k=(l-1)\beta+1}^{l\beta} \mathrm{var}[(x_k^{(j)})^2] \\
&\quad + \sum_{k=(l-1)\beta+1}^{l\beta} \sum_{\substack{m=(l-1)\beta+1 \\ k\neq m}}^{l\beta} \mathrm{cov}[(x_k^{(j)})^2, (x_m^{(j)})^2] \\
&= \beta \mathrm{var}[(x_k^{(j)})^2] \\
&\quad + \sum_{k=(l-1)\beta+1}^{l\beta} \sum_{\substack{m=(l-1)\beta+1 \\ k\neq m}}^{l\beta} \mathrm{cov}[(x_k^{(j)})^2, (x_m^{(j)})^2] \qquad (4.109)
\end{aligned}$$

2. For the case where $u \neq v$, the variance of $\tilde{w}_{j,u,v}$ is found as follows.

$$\mathrm{var}[\tilde{w}_{j,u,v}] = \mathrm{var}\left[a_l^{(u)} a_{l+j}^{(v)} \sum_{k=(l-1)\beta+1}^{l\beta} x_{k-\eta_u}^{(u)} x_{k+\beta j-\eta_v}^{(v)}\right]$$

$$\begin{aligned}
&= \sum_{k=(l-1)\beta+1}^{l\beta} \mathrm{var}[x_{k-\eta_u}^{(u)} x_{k+\beta j-\eta_v}^{(v)}] \\
&\quad + \sum_{k=(l-1)\beta+1}^{l\beta} \sum_{\substack{m=(l-1)\beta+1 \\ k\neq m}}^{l\beta} \mathrm{cov}[x_{k-\eta_u}^{(u)} x_{k+\beta j-\eta_v}^{(v)}, x_{m-\eta_u}^{(u)} x_{m+\beta j-\eta_v}^{(v)}] \\
&= \sum_{k=(l-1)\beta+1}^{l\beta} \mathrm{var}[x_{k-\eta_u}^{(u)}] \mathrm{var}[x_{k+\beta j-\eta_v}^{(v)}] \\
&\quad + \sum_{k=(l-1)\beta+1}^{l\beta} \sum_{\substack{m=(l-1)\beta+1 \\ k\neq m}}^{l\beta} \Big(\mathrm{E}[x_{k-\eta_u}^{(u)} x_{k+\beta j-\eta_v}^{(v)} x_{m-\eta_u}^{(u)} x_{m+\beta j-\eta_v}^{(v)}] \\
&\qquad - \mathrm{E}[x_{k-\eta_u}^{(u)} x_{k+\beta j-\eta_v}^{(v)}] \mathrm{E}[x_{m-\eta_u}^{(u)} x_{m+\beta j-\eta_v}^{(v)}] \Big) \\
&= \beta P_s^{(u)} P_s^{(v)}
\end{aligned}$$

$$+ \sum_{\substack{k=(l-1)\beta+1 \\ k\neq m}}^{l\beta} \sum_{m=(l-1)\beta+1}^{l\beta} \mathrm{E}[x_{k-\eta_u}^{(u)} x_{m-\eta_u}^{(u)}]\mathrm{E}[x_{k+\beta j-\eta_v}^{(v)} x_{m+\beta j-\eta_v}^{(v)}] \tag{4.110}$$

where

$$P_s^{(i)} = \mathrm{var}[x_k^{(i)}] = \mathrm{E}[(x_k^{(i)})^2] \tag{4.111}$$

denotes the average power of the ith transmitter, $i = 1, 2, \ldots, N$.

3. For the case where $u = v \neq j$, the variance of $\tilde{w}_{j,u,v}$ is found as follows.

$$\begin{aligned}
\mathrm{var}[\tilde{w}_{j,u,v}] = {} & \sum_{k=(l-1)\beta+1}^{l\beta} \mathrm{var}[x_{k-\eta_u}^{(u)} x_{k+\beta j-\eta_u}^{(u)}] \\
& + \sum_{\substack{k=(l-1)\beta+1 \\ k\neq m}}^{l\beta} \sum_{m=(l-1)\beta+1}^{l\beta} \mathrm{cov}[x_{k-\eta_u}^{(u)} x_{k+\beta j-\eta_u}^{(u)}, x_{m-\eta_u}^{(u)} x_{m+\beta j-\eta_u}^{(u)}] \\
= {} & \sum_{k=(l-1)\beta+1}^{l\beta} \left(\mathrm{E}[(x_{k-\eta_u}^{(u)} x_{k+\beta j-\eta_u}^{(u)})^2] - \mathrm{E}^2[x_{k-\eta_u}^{(u)} x_{k+\beta j-\eta_u}^{(u)}]\right) \\
& + \sum_{\substack{k=(l-1)\beta+1 \\ k\neq m}}^{l\beta} \sum_{m=(l-1)\beta+1}^{l\beta} \Big(\mathrm{E}[x_{k-\eta_u}^{(u)} x_{k+\beta j-\eta_u}^{(u)} x_{m-\eta_u}^{(u)} x_{m+\beta j-\eta_u}^{(u)}] \\
& \qquad - \mathrm{E}[x_{k-\eta_u}^{(u)} x_{k+\beta j-\eta_u}^{(u)}]\mathrm{E}[x_{m-\eta_u}^{(u)} x_{m+\beta j-\eta_u}^{(u)}]\Big)
\end{aligned} \tag{4.112}$$

4. The variances of $\tilde{p}_{j,u}$, $\tilde{q}_{j,v}$ and γ are found as follows.

$$\begin{aligned}
\mathrm{var}[\tilde{p}_{j,u}] &= \mathrm{var}\left[a_l^{(u)} \sum_{k=(l-1)\beta+1}^{l\beta} x_{k-\eta_u}^{(u)} \xi_{k+\beta j}\right] \\
&= \sum_{k=(l-1)\beta+1}^{l\beta} \mathrm{var}[x_{k-\eta_u}^{(u)} \xi_{k+\beta j}] \\
&\quad + \sum_{\substack{k=(l-1)\beta+1 \\ k\neq m}}^{l\beta} \sum_{m=(l-1)\beta+1}^{l\beta} \mathrm{cov}[x_{k-\eta_u}^{(u)} \xi_{k+\beta j}, x_{m-\eta_u}^{(u)} \xi_{m+\beta j}] \\
&= \sum_{k=(l-1)\beta+1}^{l\beta} \mathrm{var}[x_{k-\eta_u}^{(u)}]\mathrm{var}[\xi_{k+\beta j}] \\
&= \frac{\beta P_s^{(u)} N_0}{2}
\end{aligned} \tag{4.113}$$

$$\mathrm{var}[\tilde{q}_{j,v}] = \mathrm{var}\left[a_{l+j}^{(v)} \sum_{k=(l-1)\beta+1}^{l\beta} x_{k+\beta j-\eta_v}^{(v)} \xi_k\right]$$

$$
\begin{aligned}
&= \sum_{k=(l-1)\beta+1}^{l\beta} \operatorname{var}[x^{(v)}_{k+\beta j-\eta_v}\xi_k] \\
&\quad + \sum_{\substack{k=(l-1)\beta+1 \\ k\neq m}}^{l\beta} \sum_{m=(l-1)\beta+1}^{l\beta} \operatorname{cov}[x^{(v)}_{k+\beta j-\eta_v}\xi_k, x^{(v)}_{m+\beta j-\eta_v}\xi_m] \\
&= \sum_{k=(l-1)\beta+1}^{l\beta} \operatorname{var}[x^{(v)}_{k+\beta j-\eta_v}]\operatorname{var}[\xi_k] \\
&= \frac{\beta P_s^{(v)} N_0}{2} \qquad (4.114)
\end{aligned}
$$

$$
\begin{aligned}
\operatorname{var}[\gamma] &= \operatorname{var}\left[\sum_{k=(l-1)\beta+1}^{l\beta} \xi_k\xi_{k+\beta j}\right] \\
&= \sum_{k=(l-1)\beta+1}^{l\beta} \operatorname{var}[\xi_k\xi_{k+\beta j}] \\
&= \frac{\beta N_0^2}{4} \qquad (4.115)
\end{aligned}
$$

5. The covariance of $\tilde{w}_{j,u,v}$ and $\tilde{w}_{j,\tilde{u},\tilde{v}}$ for the case where $(\tilde{u},\tilde{v}) = (v,u) \neq (u,v)$ is

$$
\begin{aligned}
\operatorname{cov}[\tilde{w}_{j,u,v}, \tilde{w}_{j,\tilde{u},\tilde{v}}] &= \mathrm{E}\left[\left(a_l^{(u)} a_{l+j}^{(v)} \sum_{k=(l-1)\beta+1}^{l\beta} x^{(u)}_{k-\eta_u} x^{(v)}_{k+\beta j-\eta_v}\right)\right. \\
&\qquad \left.\times \left(a_l^{(v)} a_{l+j}^{(u)} \sum_{k=(l-1)\beta+1}^{l\beta} x^{(v)}_{k-\eta_v} x^{(u)}_{k+\beta j-\eta_u}\right)\right] \\
&= a_l^{(u)} a_{l+j}^{(u)} a_l^{(v)} a_{l+j}^{(v)} \\
&\quad \times \sum_{k=(l-1)\beta+1}^{l\beta} \sum_{m=(l-1)\beta+1}^{l\beta} \mathrm{E}[x^{(u)}_{k-\eta_u} x^{(v)}_{k+\beta j-\eta_v} x^{(v)}_{m-\eta_v} x^{(u)}_{m+\beta j-\eta_u}] \\
&= a_l^{(u)} a_{l+j}^{(u)} a_l^{(v)} a_{l+j}^{(v)} \\
&\quad \times \sum_{k=(l-1)\beta+1}^{l\beta} \sum_{m=(l-1)\beta+1}^{l\beta} \mathrm{E}[x^{(u)}_{k-\eta_u} x^{(u)}_{m+\beta j-\eta_u}]\mathrm{E}[x^{(v)}_{m-\eta_v} x^{(v)}_{k+\beta j-\eta_v}].
\end{aligned}
\qquad (4.116)
$$

Note that $u = j$ or $v = j$ is possible, but not both at the same time. When $u \neq j$, $x^{(u)}_{k-\eta_u} \neq x^{(u)}_{m+\beta j-\eta_u}$ for all possible combinations of k and m. The same condition applies when $v \neq j$.

5. Anti-Jamming Performance of Chaos-Based Digital Communication Systems Under Narrowband Sine-Wave Jammers

In the previous chapters, we have described in detail the major digital modulation schemes that exploit chaotic signals as information carriers, and discussed the analytical approaches for evaluating performance. In this and the next chapters, we focus our attention on an important performance aspect for spread-spectrum communication systems, namely, *anti-jamming capability* [Peterson *et al.* (1995)]. Intuitively, any coherent system, where the receiver knows or is able to reproduce the chaotic carriers, is expected to be considerably better in anti-jamming than its non-coherent counterpart. In chaos-based communications, coherent systems suffer serious drawbacks due to the fragility of chaos synchronization which is the main technique for reproducing the chaotic carriers [Chen and Yao (2000)]. Non-coherent systems are practically more favourable although their noise performance is inherently worse than coherent systems. It is therefore of interest to study thoroughly the anti-jamming properties of chaos-based communication systems in order to gain a better understanding of how well or poorly, in exact quantitative terms, a practical non-coherent chaotic communication system performs in comparison with the theoretically better coherent systems. Our purpose in this chapter is to present an analytical method for evaluating anti-jamming performance of chaos-based communication systems, specifically the coherent chaos-shift-keying (CSK) system [Dedieu *et al.* (1993)] and the non-coherent differential chaos-shift-keying (DCSK) system [Kolumbán *et al.* (1996)]. In particular we present results and evaluation for the case where the jamming signals are narrowband sinusoidal jammers. The case of wideband jammers will be left to the next chapter.

5.1 Systems Subject to Narrowband Sine-Wave Jammers

The basic problem we wish to investigate in this chapter is the performance of chaos-based digital communication systems when the channel is subject to a strong jamming signal as well as additive white Gaussian noise. Figure 5.1 shows a block diagram of the system under study. In our analysis, we assume that the jamming signal is a sine wave of power P_{jam} and frequency f, i.e.,

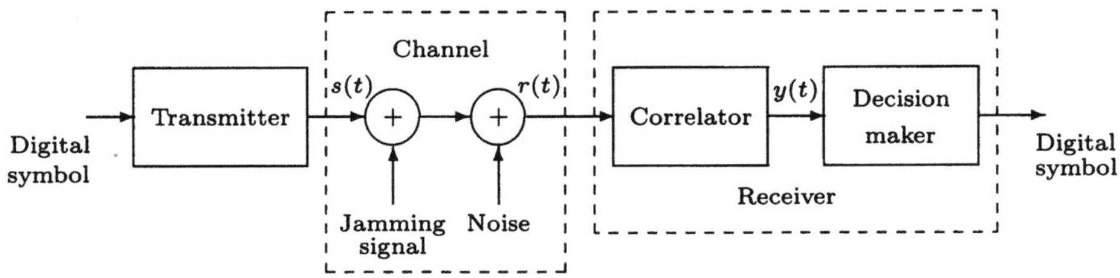

Fig. 5.1. Block diagram of a chaotic communication system with jamming and noise added.

$$s_{\text{jam}}(t) = \sqrt{2P_{\text{jam}}} \sin(2\pi ft + \theta) \tag{5.1}$$

where θ is an arbitrary constant phase angle. A sinusoidal jammer is investigated here because it is one of the most common jamming signals used and it can be generated easily. Suppose the transmitter output is $s(t)$. Assuming additive jamming and noise, the signal at the input of the receiver, $r(t)$, is given by

$$r(t) = s(t) + s_{\text{jam}}(t) + \xi(t) \tag{5.2}$$

where $\xi(t)$ is the Gaussian noise function. This signal will be used by the receiver to recover the digital symbols that have been sent to the transmitter. The method of demodulation depends on the type of the system under study. As mentioned previously, we may broadly differentiate two types of systems, namely, coherent and non-coherent systems.

In a coherent system, the receiver has the exact waveform of the chaotic signal that carries the information [Dedieu *et al.* (1993)]. This is often achieved through a chaos synchronization process. Thus, the detection method for a coherent system involves a correlation process operated upon the received signal and the known (reproduced) chaotic signal at the receiver. Specifically, if a few chaotic signals are used to represent some digital symbols, the correlation operation is able to differentiate accurately the different chaotic signals. Hence, a threshold detector can be used to recover the digital symbols. For a non-coherent system, however, no information about the chaotic signal is known to the receiver [Kolumbán *et al.* (1996); Kolumbán *et al.* (1998b)]. For example, the DCSK modulation scheme is designed such that the transmitted signal allows information to be recovered by comparing a part of the signal with another part of it.

Clearly, the coherent system should give better performance if the receiver is able to regenerate the chaotic signal and hence knows what the chaotic signal is. However, in practice (at least as known to date), there exists no robust and reliable means of reproducing chaotic signals at the receiving end. Thus, non-coherent systems still represent practical and useful alternatives for chaos-based communications. In this chapter we focus on the anti-jamming performance and compare the two types of systems. Specifically, in the following sections, we will consider the coherent CSK and the non-coherent DCSK systems and attempt to develop analytical expressions for the bit error rates of the recovered symbols.

5.2 Analysis of Anti-jamming Performance

Instead of studying the bandpass system shown in Fig. 5.1 directly, we transform the bandpass system to an equivalent baseband model such that the carrier frequency of the chaotic signal is now zero [Hasler and Schimming (2000); Proakis and Salehi (1994)]. As a consequence, the frequency of the jammer in the bandpass model is different from its equivalent in the baseband model. Based on the equivalent model, we attempt to derive the bit error rate using discrete-time analysis (see Sect. 2.6). Moreover, the chaotic, jamming and noise signals are derived from different sources. Thus, the assumption that they are independent of each other, which will be used throughout this chapter, is well justified.

5.2.1 Coherent CSK System

We consider a discrete-time binary CSK communication system, as shown in Fig. 5.2. In the transmitter, a pair of chaotic sequences, denoted by $\{\hat{x}_k\}$ and $\{\check{x}_k\}$, are generated by two chaotic maps. If the symbol "+1" is sent, $\{\hat{x}_k\}$ is transmitted during a bit period, and if "−1" is sent, $\{\check{x}_k\}$ is transmitted.

For simplicity, we consider a CSK system in which one chaos generator is used to produce chaotic signal samples $\{x_k\}$ for $k = 1, 2, \ldots$. The two possible transmitted sequences are $\{\hat{x}_k = x_k\}$ and $\{\check{x}_k = -x_k\}$. Suppose $\alpha_l \in \{-1, +1\}$ is the symbol to be sent during the lth bit period and assume that "−1" and "+1" occur with equal probabilities. Define the spreading factor, 2β, as the number of chaotic samples used to transmit one binary symbol. During the lth bit duration, i.e., for time $k = 2(l-1)\beta+1, 2(l-1)\beta+2, \ldots, 2l\beta$, the output of the transmitter is

$$s_k = \alpha_l x_k. \tag{5.3}$$

In the following analysis, we assume that a jamming signal of power P_{jam} and frequency f is added to the transmitted signal in the channel. At the same time, white Gaussian noise is also added. The "jammed" noisy signal arriving at the receiver, denoted by r_k, is given by

$$r_k = s_k + \sqrt{2P_{\text{jam}}} \sin\left(2\pi k \frac{F}{2\beta} + \theta\right) + \xi_k \tag{5.4}$$

where F is the normalized jamming frequency defined as

$$F = fT_b \tag{5.5}$$

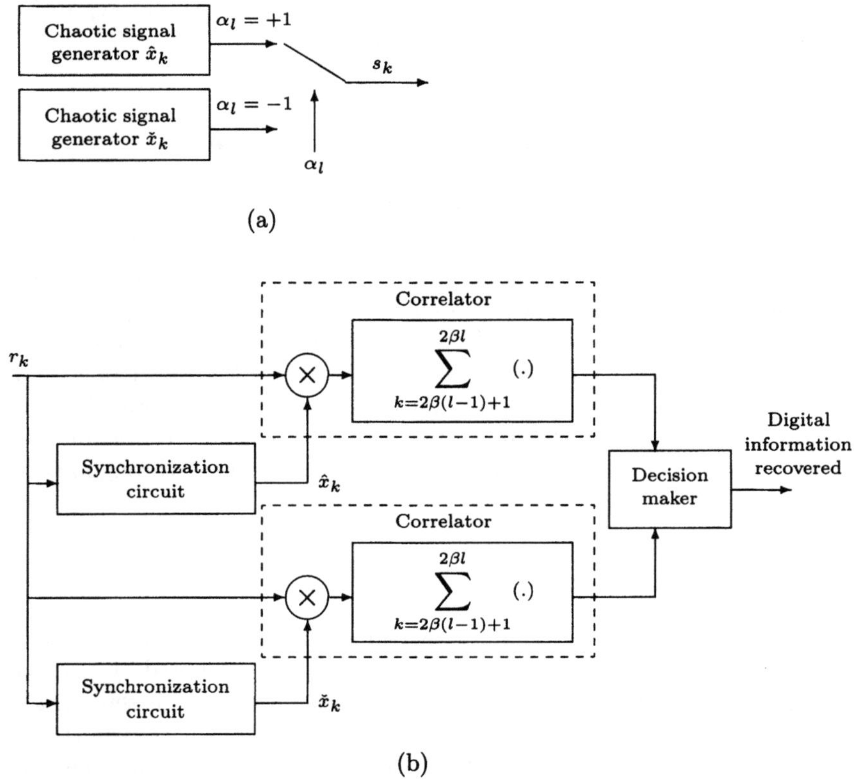

Fig. 5.2. Block diagram of a coherent CSK system. (a) Transmitter; (b) receiver.

with T_b denoting the bit duration, and θ is the initial phase angle of the jamming signal and is assumed to be an arbitrary constant angle selected from $[-\pi, \pi]$. Also, ξ_k is a Gaussian noise sample of zero mean and variance (power spectral density) $N_0/2$. We will first consider the lth bit before deriving the error probability over all transmitted bits.

Assuming that a correlator-type receiver is employed, the correlator output for the lth bit, y_l, is given by

$$\begin{aligned} y_l &= \sum_{k=2\beta(l-1)+1}^{2\beta l} r_k x_k \\ &= \underbrace{\alpha_l \sum_{k=2\beta(l-1)+1}^{2\beta l} x_k^2}_{\text{required signal}} + \underbrace{\sqrt{2P_{\text{jam}}} \sum_{k=2\beta(l-1)+1}^{2\beta l} x_k \sin\left(\frac{k\pi F}{\beta} + \theta\right)}_{\text{jamming signal}} \end{aligned}$$

$$+ \underbrace{\sum_{k=2\beta(l-1)+1}^{2\beta l} \xi_k x_k}_{\text{noise}} . \qquad (5.6)$$

Suppose a "+1" is transmitted for the lth symbol, i.e., $\alpha_l = +1$. For simplicity we write $y_l \mid (\alpha_l = +1)$ as

$$y_l \mid (\alpha_l = +1) = A + B + C \qquad (5.7)$$

where A, B and C are the required signal, jamming signal and noise, respectively, and are defined as

$$A = \sum_{k=2\beta(l-1)+1}^{2\beta l} x_k^2 \qquad (5.8)$$

$$B = \sqrt{2P_{\text{jam}}} \sum_{k=2\beta(l-1)+1}^{2\beta l} x_k \sin\left(\frac{k\pi F}{\beta} + \theta\right) \qquad (5.9)$$

$$C = \sum_{k=2\beta(l-1)+1}^{2\beta l} \xi_k x_k . \qquad (5.10)$$

The mean of $y_l \mid (\alpha_l = +1)$ is

$$\begin{aligned}
\mathrm{E}[y_l \mid (\alpha_l = +1)] &= \mathrm{E}[A] + \mathrm{E}[B] + \mathrm{E}[C] \\
&= \mathrm{E}\left[\sum_{k=2\beta(l-1)+1}^{2\beta l} x_k^2\right] \\
&\quad + \mathrm{E}\left[\sqrt{2P_{\text{jam}}} \sum_{k=2\beta(l-1)+1}^{2\beta l} x_k \sin\left(\frac{k\pi F}{\beta} + \theta\right)\right] \\
&\quad + \mathrm{E}\left[\sum_{k=2\beta(l-1)+1}^{2\beta l} \xi_k x_k\right] \\
&= \sum_{k=2\beta(l-1)+1}^{2\beta l} \mathrm{E}[x_k^2] \\
&\quad + \sqrt{2P_{\text{jam}}} \sum_{k=2\beta(l-1)+1}^{2\beta l} \mathrm{E}[x_k]\mathrm{E}\left[\sin\left(\frac{k\pi F}{\beta} + \theta\right)\right] \\
&\quad + \sum_{k=2\beta(l-1)+1}^{2\beta l} \mathrm{E}[\xi_k]\mathrm{E}[x_k] \\
&= 2\beta P_s + 2\beta\sqrt{2P_{\text{jam}}}\mathrm{E}[x_k]\mathrm{E}\left[\sin\left(\frac{k\pi F}{\beta} + \theta\right)\right] \qquad (5.11)
\end{aligned}$$

where $P_s = \mathrm{E}[x_k^2]$ denotes the average power of the chaotic signal. The last equality holds because $\mathrm{E}[\xi_k] = 0$. The variance of $y_l \mid (\alpha_l = +1)$ is

$$\begin{aligned}\mathrm{var}[y_l \mid (\alpha_l = +1)] &= \mathrm{var}[A] + \mathrm{var}[B] + \mathrm{var}[C] + 2\mathrm{cov}[A, B] \\ &\quad + 2\mathrm{cov}[B, C] + 2\mathrm{cov}[A, C]\end{aligned} \tag{5.12}$$

where $\mathrm{cov}[X, Y]$ is the covariance between X and Y defined as

$$\mathrm{cov}[X, Y] = \mathrm{E}[XY] - \mathrm{E}[X]\mathrm{E}[Y]. \tag{5.13}$$

It can be readily shown that both $\mathrm{cov}[A, C]$ and $\mathrm{cov}[B, C]$ are zero. Hence, (5.12) is simplified to

$$\mathrm{var}[y_l \mid (\alpha_l = +1)] = \mathrm{var}[A] + \mathrm{var}[B] + \mathrm{var}[C] + 2\mathrm{cov}[A, B]. \tag{5.14}$$

The average power of the chaotic signal, denoted by P_s, can be computed by numerical simulation, or by numerical integration if the probability density function of $\{x_k\}$ is available. The second term on the right hand side of (5.11), and the variance and covariance terms in (5.14) can also be computed by numerical simulation. Hence, $\mathrm{E}[y_l \mid (\alpha_l = +1)]$ and $\mathrm{var}[y_l \mid (\alpha_l = +1)]$ can be evaluated.

For the lth symbol, an error occurs when $y_l \le 0 \mid (\alpha_l = +1)$. Since $y_l \mid (\alpha_l = +1)$ is the sum of a large number of random variables, we may assume that it follows a normal distribution. The error probability is thus given by

$$\mathrm{Prob}(y_l \le 0 \mid (\alpha_l = +1)) = \frac{1}{2}\mathrm{erfc}\left(\frac{\mathrm{E}\left[y_l \mid (\alpha_l = +1)\right]}{\sqrt{2\mathrm{var}\left[y_l \mid (\alpha_l = +1)\right]}}\right) \tag{5.15}$$

where the complementary error function, erfc(.), is defined as

$$\mathrm{erfc}(\psi) \equiv \frac{2}{\sqrt{\pi}} \int_{\psi}^{\infty} e^{-\lambda^2} d\lambda. \tag{5.16}$$

Similarly, when a "−1" is sent, the output of the correlator can be shown equal to

$$y_l \mid (\alpha_l = -1) = -A + B + C. \tag{5.17}$$

Likewise, the mean and variance of $y_l \mid (\alpha_l = -1)$, denoted by $\mathrm{E}[y_l \mid (\alpha_l = -1)]$ and $\mathrm{var}[y_l \mid (\alpha_l = -1)]$, can be derived, i.e.,

$$\begin{aligned}\mathrm{E}[y_l \mid (\alpha_l = -1)] &= -2\beta P_s \\ &\quad + 2\beta\sqrt{2P_{\mathrm{jam}}}\mathrm{E}[x_k]\mathrm{E}\left[\sin\left(\frac{k\pi F}{\beta} + \theta\right)\right] \qquad (5.18) \\ \mathrm{var}[y_l \mid (\alpha_l = -1)] &= \mathrm{var}[A] + \mathrm{var}[B] + \mathrm{var}[C] - 2\mathrm{cov}[A, B] \qquad (5.19)\end{aligned}$$

where A, B, and C are defined as in (5.8) to (5.10). The corresponding error probability is

$$\operatorname{Prob}(y_l > 0 \mid (\alpha_l = -1)) = \frac{1}{2}\operatorname{erfc}\left(\frac{-\mathrm{E}\left[y_l \mid (\alpha_l = -1)\right]}{\sqrt{2\operatorname{var}\left[y_l \mid (\alpha_l = -1)\right]}}\right). \tag{5.20}$$

Hence, the error probability of the lth transmitted symbol is

$$\begin{aligned}
\mathrm{BER}_{\mathrm{CSK}}^{(l)} &= \operatorname{Prob}(\alpha_l = 1) \times \operatorname{Prob}(y_l \leq 0 \mid \alpha_l = 1) \\
&\quad + \operatorname{Prob}(\alpha_l = -1) \times \operatorname{Prob}(y_l > 0 \mid \alpha_l = -1) \\
&= \frac{1}{4}\left[\operatorname{erfc}\left(\frac{\mathrm{E}\left[y_l \mid (\alpha_l = +1)\right]}{\sqrt{2\operatorname{var}\left[y_l \mid (\alpha_l = +1)\right]}}\right)\right. \\
&\quad \left. + \operatorname{erfc}\left(\frac{-\mathrm{E}\left[y_l \mid (\alpha_l = -1)\right]}{\sqrt{2\operatorname{var}\left[y_l \mid (\alpha_l = -1)\right]}}\right)\right].
\end{aligned} \tag{5.21}$$

Since $\mathrm{E}\left[y_l \mid (\alpha_l = +1)\right]$, $\operatorname{var}[y_l \mid (\alpha_l = +1)]$, $\mathrm{E}\left[y_l \mid (\alpha_l = -1)\right]$ and $\operatorname{var}[y_l \mid (\alpha_l = -1)]$ (given by (5.11), (5.14), (5.18) and (5.19) respectively) are all independent of l, $\mathrm{BER}_{\mathrm{CSK}}^{(l)}$ is also independent of l. Thus, the bit error probability of the lth transmitted symbol is the same as the bit error rate (BER) of the whole system.The BER of the noisy CSK system under the influence of a sinusoidal jammer, denoted by $\mathrm{BER}_{\mathrm{CSK}}$, is therefore

$$\begin{aligned}
\mathrm{BER}_{\mathrm{CSK}} = \mathrm{BER}_{\mathrm{CSK}}^{(l)} &= \frac{1}{4}\left[\operatorname{erfc}\left(\frac{\mathrm{E}\left[y_l \mid (\alpha_l = +1)\right]}{\sqrt{2\operatorname{var}\left[y_l \mid (\alpha_l = +1)\right]}}\right)\right. \\
&\quad \left. + \operatorname{erfc}\left(\frac{-\mathrm{E}\left[y_l \mid (\alpha_l = -1)\right]}{\sqrt{2\operatorname{var}\left[y_l \mid (\alpha_l = -1)\right]}}\right)\right].
\end{aligned} \tag{5.22}$$

Hence, (5.11), (5.14), (5.18) and (5.19) can be computed and substituted into (5.22) to obtain the BER of the system.

Further simplifications of the equations are possible if several assumptions are made.

1. The mean value of $\{x_k\}$ is zero. The justification for this assumption is that no power should be wasted in sending non-information-bearing dc component through the channel. In practice, any dc component generated by the chaos generator can be removed artificially before transmission.
2. $\{x_k\}$ has a vanishing autovariance function. This can be proved to be true for some chaotic sequences.
3. $\{x_k^2\}$ has a vanishing autovariance function.
4. The normalized frequency F is an irrational number because the jamming signal and the chaotic signals are derived from different sources which are not synchronized. Therefore, we can assume that the sampled sinusoidal signal $\sin\left(\frac{k\pi F}{\beta} + \theta\right)$ is not periodic.

From the above assumptions, we have

$$\mathrm{E}[x_k] = 0 \tag{5.23}$$
$$\mathrm{cov}[x_k, x_n] = \mathrm{E}[x_k x_n] - \mathrm{E}[x_k]\mathrm{E}[x_n] = 0 \quad \text{for} \quad k \neq n \tag{5.24}$$
$$\mathrm{cov}[x_k^2, x_n^2] = \mathrm{E}[x_k^2 x_n^2] - \mathrm{E}[x_k^2]\mathrm{E}[x_n^2] = 0 \quad \text{for} \quad k \neq n \tag{5.25}$$
$$\mathrm{E}\left[\sin\left(\frac{k\pi F}{\beta} + \theta\right)\right] = 0. \tag{5.26}$$

Thus, (5.11) and (5.18) are simplified to

$$\mathrm{E}[y_l \mid (\alpha_l = +1)] = 2\beta P_s \tag{5.27}$$
$$\mathrm{E}[y_l \mid (\alpha_l = -1)] = -2\beta P_s \tag{5.28}$$

and the variances of the variables A, B and C, and the covariance between A and B are given by

$$\mathrm{var}[A] = 2\beta\Lambda \tag{5.29}$$
$$\mathrm{var}[B] = 2\beta P_{\mathrm{jam}} P_s \tag{5.30}$$
$$\mathrm{var}[C] = \beta N_0 P_s \tag{5.31}$$
$$\mathrm{cov}[A, B] = 0 \tag{5.32}$$

where Λ is the variance of $\{x_k^2\}$, i.e.,

$$\Lambda = \mathrm{var}[x_k^2]. \tag{5.33}$$

Hence, both (5.14) and (5.19) are given by

$$\begin{aligned}\mathrm{var}[y_l \mid (\alpha_l = +1)] &= \mathrm{var}[y_l \mid (\alpha_l = -1)] \\ &= 2\beta\Lambda + 2\beta P_{\mathrm{jam}} P_s + \beta N_0 P_s \\ &= \beta(2\Lambda + 2P_{\mathrm{jam}} P_s + N_0 P_s).\end{aligned} \tag{5.34}$$

Note that the larger the signal power P_s, the larger the variance of the correlator output y_l. Substituting (5.27), (5.28) and (5.34) into (5.22), the BER can be found as

$$\begin{aligned}\mathrm{BER_{CSK}} &= \frac{1}{2}\mathrm{erfc}\left(\frac{2\beta P_s}{\sqrt{2\beta(2\Lambda + 2P_{\mathrm{jam}} P_s + N_0 P_s)}}\right) \\ &= \frac{1}{2}\mathrm{erfc}\left[\left(\frac{\Psi}{\beta} + \frac{P_{\mathrm{jam}}}{\beta P_s} + \frac{N_0}{2\beta P_s}\right)^{-\frac{1}{2}}\right] \qquad (5.35) \\ &= \frac{1}{2}\mathrm{erfc}\left[\left(\frac{\Psi}{\beta} + \left(\frac{E_b}{2P_{\mathrm{jam}}}\right)^{-1} + \left(\frac{E_b}{N_0}\right)^{-1}\right)^{-\frac{1}{2}}\right] \qquad (5.36)\end{aligned}$$

where $E_b = 2\beta P_s$ denotes the average bit energy and $\Psi = \mathrm{var}[x_k^2]/P_s^2$ is a constant depending upon the type of the chaotic sequence used (see Sect. 3.2.3.2). The expressions given in (5.35) and (5.36) are thus the analytical BER for the jammed noisy coherent CSK system. Note that for fixed jamming power P_{jam} and noise power spectral density $N_0/2$, the BER can be improved by making one or more of the following adjustments.

1. Reduce the variance of $\{x_k^2\}$ for any given P_s.
2. Increase the spreading factor 2β.
3. Increase the bit energy E_b.

When the jamming power is zero, i.e., $P_{\text{jam}} = 0$, it can be readily shown that the BER reduces to

$$\text{BER}_{\text{CSK}} \mid (P_{\text{jam}} = 0) = \frac{1}{2}\text{erfc}\left[\left(\frac{\Psi}{\beta} + \frac{N_0}{2\beta P_s}\right)^{-\frac{1}{2}}\right] \quad (5.37)$$

$$= \frac{1}{2}\text{erfc}\left[\left(\frac{\Psi}{\beta} + \left(\frac{E_b}{N_0}\right)^{-1}\right)^{-\frac{1}{2}}\right]. \quad (5.38)$$

Note that in the derivation of the BER, it has been assumed that the exact reference is available in the receiver. While the re-construction of the exact reference at the receiver, especially in the presence of noise and jamming, remains an unsolved problem at present, the BER derived can nonetheless provide performance benchmarks for evaluating other chaotic modulation/demodulation techniques under similar jamming conditions.

Example — Consider the case where the logistic map is used for chaos generation. The form of the map is

$$x_{k+1} = g(x_k) = 1 - 2x_k^2 \quad (5.39)$$

and the invariant probability density function of $\{x_k\}$, denoted by $\rho(x)$, is [Kohda and Tsuneda (1994)]

$$\rho(x) = \begin{cases} \dfrac{1}{\pi\sqrt{1-x^2}} & \text{if } |x| < 1 \\ 0 & \text{otherwise.} \end{cases} \quad (5.40)$$

It can be shown easily that the mean value of $\{x_k\}$ is zero and $\{x_k\}$ has a vanishing autovariance function. Also, it is proved in Appendix 3B that the autovariance for $\{x_k^2\}$ is vanishing for the logistic map. In other words, the first three assumptions made earlier are justified. Also, we have

$$P_s = \text{E}[x_k^2] = \int_{-\infty}^{\infty} x^2\rho(x)dx = \int_{-1}^{1} x^2\rho(x)dx = \frac{1}{2} \quad (5.41)$$

$$\Lambda = \text{var}[x_k^2] = \text{E}[x_k^4] - \text{E}^2[x_k^2] = \int_{-1}^{1} x^4\rho(x)dx - \frac{1}{4} = \frac{1}{8}. \quad (5.42)$$

For the case where the logistic map is used to generate the chaotic samples, we substitute (5.41) and (5.42) into (5.35) to obtain the BER, i.e.,

$$\text{BER}_{\text{CSK}} = \frac{1}{2}\text{erfc}\left(\sqrt{\frac{2\beta}{1 + 4P_{\text{jam}} + 2N_0}}\right). \quad (5.43)$$

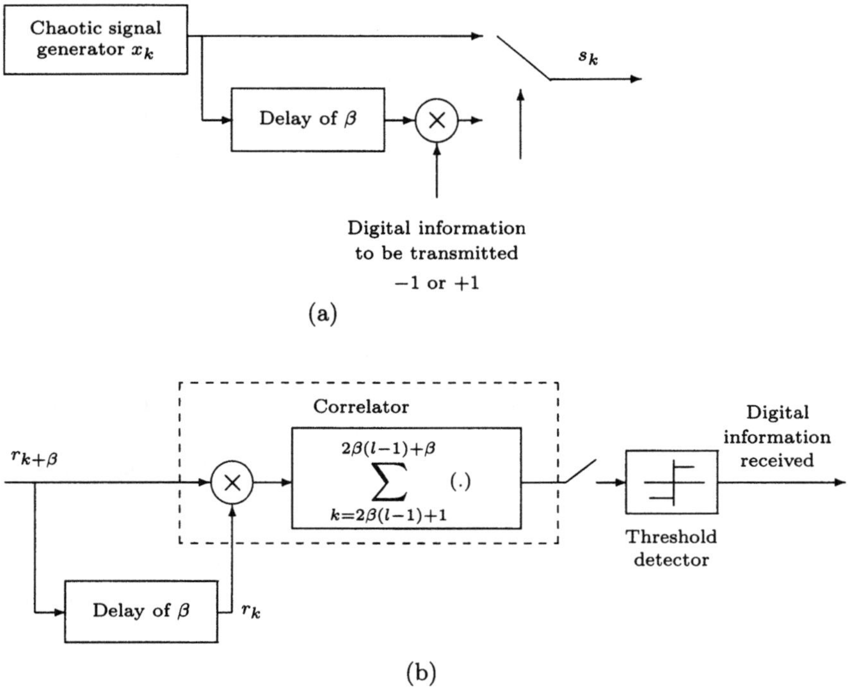

Fig. 5.3. Block diagram of a non-coherent DCSK system. (a) Transmitter; (b) receiver.

5.2.2 Non-Coherent DCSK System

In this section, we consider the DCSK system. The basic modulation process involves dividing the bit period into two equal slots. The first slot carries a reference chaotic signal, and the second slot bears the information. For a binary system, the second slot is the same copy or an inverted copy of the first slot depending upon the symbol sent being "+1" or "−1". This structural arrangement allows the detection to be done in a non-coherent manner requiring no reproduction of the same chaotic carrying signals at the receiver. Essentially, the detection of a DCSK signal can be accomplished by correlating the first and the second slots of the same symbol and comparing the correlator output with a threshold. Figure 5.3 shows the block diagrams of a DCSK transmitter and receiver pair.

As in the previous case, we assume that the signal is jammed additively by a sine wave of power P_{jam} and normalized frequency F. In addition, the received signal contains additive white Gaussian noise. Using the same notations and functions as defined in Sect. 5.2.1, the signal transmitted in the lth bit can be written as

$$s_k = \begin{cases} x_k & \text{for } k = 2\beta(l-1)+1, 2\beta(l-1)+2, \ldots, 2\beta(l-1)+\beta \\ \alpha_l x_{k-\beta} & \text{for } k = 2\beta(l-1)+\beta+1, 2\beta(l-1)+\beta+2, \ldots, 2\beta l \end{cases} \tag{5.44}$$

where α_l takes on the values of ± 1 according to the bit being sent. The noisy jammed signal, r_k, within a bit duration is given by

$$r_k = s_k + \sqrt{2P_{\text{jam}}} \sin\left(\frac{k\pi F}{\beta} + \theta\right) + \xi_k \tag{5.45}$$

where the symbols F, θ and ξ_k are as defined previously in Sect. 5.2.1.

At the receiver, the detector essentially calculates the correlation of the reference and data slots of the same symbol. We consider the output of the correlator for the lth received bit, y_l, which is given by

$$\begin{aligned}
y_l &= \sum_{k=2\beta(l-1)+1}^{2\beta(l-1)+\beta} r_k r_{k+\beta} \\
&= \sum_{k=2\beta(l-1)+1}^{2\beta(l-1)+\beta} \left[x_k + \sqrt{2P_{\text{jam}}} \sin\left(\frac{k\pi F}{\beta} + \theta\right) + \xi_k\right] \\
&\quad \times \left[\alpha_l x_k + \sqrt{2P_{\text{jam}}} \sin\left(\frac{(k+\beta)\pi F}{\beta} + \theta\right) + \xi_{k+\beta}\right] \\
&= \alpha_l \sum_{k=2\beta(l-1)+1}^{2\beta(l-1)+\beta} x_k^2 + \sqrt{2P_{\text{jam}}} \\
&\quad \times \sum_{k=2\beta(l-1)+1}^{2\beta(l-1)+\beta} x_k \left[\sin\left(\frac{(k+\beta)\pi F}{\beta} + \theta\right) + \alpha_l \sin\left(\frac{k\pi F}{\beta} + \theta\right)\right] \\
&\quad + 2P_{\text{jam}} \sum_{k=2\beta(l-1)+1}^{2\beta(l-1)+\beta} \sin\left(\frac{(k+\beta)\pi F}{\beta} + \theta\right) \sin\left(\frac{k\pi F}{\beta} + \theta\right) \\
&\quad + \sum_{k=2\beta(l-1)+1}^{2\beta(l-1)+\beta} x_k(\xi_{k+\beta} + \alpha_l \xi_k) \\
&\quad + \sqrt{2P_{\text{jam}}} \sum_{k=2\beta(l-1)+1}^{2\beta(l-1)+\beta} \left[\xi_{k+\beta} \sin\left(\frac{k\pi F}{\beta} + \theta\right)\right. \\
&\quad \left. + \xi_k \sin\left(\frac{(k+\beta)\pi F}{\beta} + \theta\right)\right] + \sum_{k=2\beta(l-1)+1}^{2\beta(l-1)+\beta} \xi_k \xi_{k+\beta}.
\end{aligned} \tag{5.46}$$

Define

$$A = \sum_{k=2\beta(l-1)+1}^{2\beta(l-1)+\beta} x_k^2 \tag{5.47}$$

$$B = 2\sqrt{2P_{\text{jam}}}\cos\left(\frac{\pi F}{2}\right)\sum_{k=2\beta(l-1)+1}^{2\beta(l-1)+\beta} x_k \sin\left(\frac{k\pi F}{\beta}+\frac{\pi F}{2}+\theta\right) \quad (5.48)$$

$$B' = 2\sqrt{2P_{\text{jam}}}\sin\left(\frac{\pi F}{2}\right)\sum_{k=2\beta(l-1)+1}^{2\beta(l-1)+\beta} x_k \cos\left(\frac{k\pi F}{\beta}+\frac{\pi F}{2}+\theta\right) \quad (5.49)$$

$$C = \beta P_{\text{jam}}\cos(\pi F) \quad (5.50)$$

$$D = \frac{-P_{\text{jam}}\sin(\pi F)}{\sin(\pi F/\beta)}\cos\left(4l\pi F - 2\pi F + \frac{\pi F}{\beta} + 2\theta\right) \quad (5.51)$$

$$G = \sum_{k=2\beta(l-1)+1}^{2\beta(l-1)+\beta} x_k \xi_{k+\beta} \quad (5.52)$$

$$H = \sum_{k=2\beta(l-1)+1}^{2\beta(l-1)+\beta} x_k \xi_k \quad (5.53)$$

$$J = \sqrt{2P_{\text{jam}}}\sum_{k=2\beta(l-1)+1}^{2\beta(l-1)+\beta} \sin\left(\frac{k\pi F}{\beta}+\theta\right)\xi_{k+\beta} \quad (5.54)$$

$$K = \sqrt{2P_{\text{jam}}}\sum_{k=2\beta(l-1)+1}^{2\beta(l-1)+\beta} \sin\left(\frac{(k+\beta)\pi F}{\beta}+\theta\right)\xi_k \quad (5.55)$$

$$L = \sum_{k=2\beta(l-1)+1}^{2\beta(l-1)+\beta} \xi_k \xi_{k+\beta}. \quad (5.56)$$

Using (5.47) to (5.56), it can be easily shown that when $\alpha_l = +1$ and $\alpha_l = -1$, (5.46) becomes, respectively,

$$y_l \mid (\alpha_l = +1) = \underbrace{A}_{\text{required signal}} + \underbrace{B + C + D}_{\text{jamming signal}} + \underbrace{G + H + J + K + L}_{\text{noise}} \quad (5.57)$$

and

$$y_l \mid (\alpha_l = -1) = \underbrace{-A}_{\text{required signal}} + \underbrace{B' + C + D}_{\text{jamming signal}} + \underbrace{G - H + J + K + L}_{\text{noise}}. \quad (5.58)$$

Similar to the case of CSK, the means and variances of y_l given a "+1" or "−1" is sent can be evaluated by numerical simulations. Denote the respective means and variances by $\mathrm{E}[y_l \mid (\alpha_l = \pm 1)]$ and $\mathrm{var}[y_l \mid (\alpha_l = \pm 1)]$. Since $y_l \mid (\alpha_l = \pm 1)$ is the sum of a large number of random variables, we may assume that it is normally distributed. Hence, the approximate error probability of the lth transmitted bit is

$$\mathrm{BER}_{\mathrm{DCSK}}^{(l)} = \mathrm{Prob}(\alpha_l = 1) \times \mathrm{Prob}(y_l \leq 0 \mid (\alpha_l = +1))$$

$$
\begin{aligned}
&+ \operatorname{Prob}(\alpha_l = -1) \times \operatorname{Prob}(y_l > 0 \mid (\alpha_l = -1)) \\
= \ & \frac{1}{4}\left[\operatorname{erfc}\left(\frac{\mathrm{E}\left[y_l \mid (\alpha_l = +1)\right]}{\sqrt{2\operatorname{var}\left[y_l \mid (\alpha_l = +1)\right]}}\right)\right. \\
&\left. + \operatorname{erfc}\left(\frac{-\mathrm{E}\left[y_l \mid (\alpha_l = -1)\right]}{\sqrt{2\operatorname{var}\left[y_l \mid (\alpha_l = -1)\right]}}\right)\right]. \qquad (5.59)
\end{aligned}
$$

Note that the approximation gets better when the spreading factor becomes larger.

To simplify the analysis, we make similar assumptions as in Sect. 5.2.1, i.e.,

1. The mean value of $\{x_k\}$ is zero.
2. $\{x_k\}$ has a vanishing autovariance function.
3. $\{x_k^2\}$ has a vanishing autovariance function.
4. The normalized frequency F is an irrational number.

With these assumptions, it can be shown that

$$
\begin{aligned}
\mathrm{E}\left[y_l \mid (\alpha_l = +1)\right] &= \mathrm{E}[A] + \mathrm{E}[B] + \mathrm{E}[C] + \mathrm{E}[D] + \mathrm{E}[G] \\
&\quad + \mathrm{E}[H] + \mathrm{E}[J] + \mathrm{E}[K] + \mathrm{E}[L] \qquad (5.60) \\
\mathrm{E}\left[y_l \mid (\alpha_l = -1)\right] &= -\mathrm{E}[A] + \mathrm{E}[B'] + \mathrm{E}[C] + \mathrm{E}[D] + \mathrm{E}[G] \\
&\quad - \mathrm{E}[H] + \mathrm{E}[J] + \mathrm{E}[K] + \mathrm{E}[L] \qquad (5.61) \\
\operatorname{var}\left[y_l \mid (\alpha_l = +1)\right] &\approx \operatorname{var}[A] + \operatorname{var}[B] + \operatorname{var}[C] + \operatorname{var}[D] \\
&\quad + \operatorname{var}[G] + \operatorname{var}[H] + \operatorname{var}[J] \\
&\quad + \operatorname{var}[K] + \operatorname{var}[L] \qquad (5.62) \\
\operatorname{var}\left[y_l \mid (\alpha_l = -1)\right] &\approx \operatorname{var}[A] + \operatorname{var}[B'] + \operatorname{var}[C] + \operatorname{var}[D] \\
&\quad + \operatorname{var}[G] + \operatorname{var}[H] + \operatorname{var}[J] \\
&\quad + \operatorname{var}[K] + \operatorname{var}[L]. \qquad (5.63)
\end{aligned}
$$

Moreover, it can be readily shown that the relevant means and variances are

$$
\begin{array}{ll}
\mathrm{E}[A] = \beta E[x_k^2] \equiv \beta P_s & \operatorname{var}[A] = \beta \operatorname{var}[x_k^2] \equiv \beta \Lambda \\
\mathrm{E}[B] = 0 & \operatorname{var}[B] = 4\beta P_s P_{\text{jam}} \cos^2\left(\frac{\pi F}{2}\right) \\
\mathrm{E}[B'] = 0 & \operatorname{var}[B'] = 4\beta P_s P_{\text{jam}} \sin^2\left(\frac{\pi F}{2}\right) \\
\mathrm{E}[C] = \beta P_{\text{jam}} \cos(\pi F) & \operatorname{var}[C] = 0 \\
\mathrm{E}[D] = \dfrac{-P_{\text{jam}} \sin(\pi F)}{\sin(\pi F/\beta)} & \operatorname{var}[D] = 0 \\
\qquad \times \cos\left(4l\pi F - 2\pi F\right. & \qquad\qquad\qquad\qquad (5.64) \\
\qquad \left. + \frac{\pi F}{\beta} + 2\theta\right) & \\
\mathrm{E}[G] = 0 & \operatorname{var}[G] = \beta P_s N_0/2 \\
\mathrm{E}[H] = 0 & \operatorname{var}[H] = \beta P_s N_0/2 \\
\mathrm{E}[J] = 0 & \operatorname{var}[J] = \beta P_{\text{jam}} N_0/2 \\
\mathrm{E}[K] = 0 & \operatorname{var}[K] = \beta P_{\text{jam}} N_0/2 \\
\mathrm{E}[L] = 0 & \operatorname{var}[L] = \beta N_0^2/4.
\end{array}
$$

Substituting (5.64) into (5.60)–(5.63), we obtain

$$\begin{aligned} \mathrm{E}\left[y_l \mid (\alpha_l = +1)\right] &= \beta P_s + \beta P_{\text{jam}} \cos(\pi F) \\ &\quad - \frac{P_{\text{jam}} \sin(\pi F)}{\sin(\pi F/\beta)} \\ &\quad \times \cos\left(4l\pi F - 2\pi F + \frac{\pi F}{\beta} + 2\theta\right) \end{aligned} \quad (5.65)$$

$$\begin{aligned} \mathrm{E}\left[y_l \mid (\alpha_l = -1)\right] &= -\beta P_s + \beta P_{\text{jam}} \cos(\pi F) \\ &\quad - \frac{P_{\text{jam}} \sin(\pi F)}{\sin(\pi F/\beta)} \\ &\quad \times \cos\left(4l\pi F - 2\pi F + \frac{\pi F}{\beta} + 2\theta\right) \end{aligned} \quad (5.66)$$

$$\begin{aligned} \operatorname{var}\left[y_l \mid (\alpha_l = +1)\right] &\approx \beta \Lambda + 4\beta P_s P_{\text{jam}} \cos^2\left(\frac{\pi F}{2}\right) \\ &\quad + \beta P_s N_0 + \beta P_{\text{jam}} N_0 + \frac{\beta N_0^2}{4} \end{aligned} \quad (5.67)$$

$$\begin{aligned} \operatorname{var}\left[y_l \mid (\alpha_l = -1)\right] &\approx \beta \Lambda + 4\beta P_s P_{\text{jam}} \sin^2\left(\frac{\pi F}{2}\right) \\ &\quad + \beta P_s N_0 + \beta P_{\text{jam}} N_0 + \frac{\beta N_0^2}{4}. \end{aligned} \quad (5.68)$$

Thus, (5.59) now becomes

$$\begin{aligned} &\mathrm{BER}_{\mathrm{DCSK}}^{(l)} \\ &= \frac{1}{4}\left[\operatorname{erfc}\left(\frac{\beta(P_s + P_{\text{jam}}\cos(\pi F)) - \frac{P_{\text{jam}}\sin(\pi F)}{\sin(\pi F/\beta)}\cos\left(F(4l\pi - 2\pi + \frac{\pi}{\beta}) + 2\theta\right)}{\sqrt{2\beta\Lambda + 8\beta P_s P_{\text{jam}}\cos^2\left(\frac{\pi F}{2}\right) + 2\beta P_s N_0 + 2\beta P_{\text{jam}} N_0 + \frac{\beta N_0^2}{2}}}\right)\right. \\ &\quad \left. + \operatorname{erfc}\left(\frac{\beta(P_s - P_{\text{jam}}\cos(\pi F)) + \frac{P_{\text{jam}}\sin(\pi F)}{\sin(\pi F/\beta)}\cos\left(F(4l\pi - 2\pi + \frac{\pi}{\beta}) + 2\theta\right)}{\sqrt{2\beta\Lambda + 8\beta P_s P_{\text{jam}}\sin^2\left(\frac{\pi F}{2}\right) + 2\beta P_s N_0 + 2\beta P_{\text{jam}} N_0 + \frac{\beta N_0^2}{2}}}\right)\right] \end{aligned} \quad (5.69)$$

The average error rate over all transmitted bits is thus

$$\mathrm{BER}_{\mathrm{DCSK}} = \lim_{N_s \to \infty} \frac{1}{N_s} \sum_{l=1}^{N_s} \mathrm{BER}_{\mathrm{DCSK}}^{(l)}. \quad (5.70)$$

In (5.69), only the term $\cos\left(4l\pi F - 2\pi F + \frac{\pi F}{\beta} + 2\theta\right)$ is a function of l. Moreover, it has been assumed that F is irrational. Therefore, as l varies, $4l\pi F$ and hence $\left(4l\pi F - 2\pi F + \frac{\pi F}{\beta} + 2\theta\right)$ will be uniformly distributed in $[-\pi, +\pi]$. Denoting $4l\pi F - 2\pi F + \frac{\pi F}{\beta} + 2\theta$ by γ, (5.70) can be rewritten as

$$\mathrm{BER}_{\mathrm{DCSK}} = \int_{-\pi}^{+\pi} \mathrm{BER}_{\mathrm{DCSK}}^{(l)}\, \nu(\gamma)\, d\gamma \quad (5.71)$$

where $\nu(\gamma) = \frac{1}{2\pi}$ is the probability density function of γ. Define $W = \cos\gamma$. The probability density function of W, denoted by $\mu(W)$, is

$$\mu(W) = \begin{cases} \dfrac{1}{\pi\sqrt{1-W^2}} & \text{for } |W| \le 1 \\ 0 & \text{otherwise.} \end{cases} \tag{5.72}$$

Thus, (5.71) becomes

$$\begin{aligned}
\mathrm{BER_{DCSK}} &= \int_{-1}^{+1} \mathrm{BER}^{(l)}_{\mathrm{DCSK}}\, \mu(W)\, dW \\
&= \frac{1}{4}\int_{-1}^{+1} \left[\mathrm{erfc}\left(\frac{\beta P_s + \beta P_{\mathrm{jam}}\cos(\pi F) - \frac{P_{\mathrm{jam}}\sin(\pi F)}{\sin(\pi F/\beta)} W}{\sqrt{2\beta\Lambda + 8\beta P_s P_{\mathrm{jam}}\cos^2\left(\frac{\pi F}{2}\right) + 2\beta P_s N_0 + 2\beta P_{\mathrm{jam}} N_0 + \frac{\beta N_0^2}{2}}} \right) \right. \\
&\quad \left. + \,\mathrm{erfc}\left(\frac{\beta P_s - \beta P_{\mathrm{jam}}\cos(\pi F) + \frac{P_{\mathrm{jam}}\sin(\pi F)}{\sin(\pi F/\beta)} W}{\sqrt{2\beta\Lambda + 8\beta P_s P_{\mathrm{jam}}\sin^2\left(\frac{\pi F}{2}\right) + 2\beta P_s N_0 + 2\beta P_{\mathrm{jam}} N_0 + \frac{\beta N_0^2}{2}}} \right) \right] \\
&\quad \times \frac{1}{\pi\sqrt{1-W^2}}\, dW
\end{aligned} \tag{5.73}$$

$$\begin{aligned}
&= \frac{1}{4}\int_{-1}^{+1} \left[\mathrm{erfc}\left(\frac{\frac{E_b}{2} + \beta P_{\mathrm{jam}}\cos(\pi F) - \frac{P_{\mathrm{jam}}\sin(\pi F)}{\sin(\pi F/\beta)} W}{\sqrt{2\beta\Lambda + 4E_b P_{\mathrm{jam}}\cos^2\left(\frac{\pi F}{2}\right) + E_b N_0 + 2\beta P_{\mathrm{jam}} N_0 + \frac{\beta N_0^2}{2}}} \right) \right. \\
&\quad \left. + \,\mathrm{erfc}\left(\frac{\frac{E_b}{2} - \beta P_{\mathrm{jam}}\cos(\pi F) + \frac{P_{\mathrm{jam}}\sin(\pi F)}{\sin(\pi F/\beta)} W}{\sqrt{2\beta\Lambda + 4E_b P_{\mathrm{jam}}\sin^2\left(\frac{\pi F}{2}\right) + E_b N_0 + 2\beta P_{\mathrm{jam}} N_0 + \frac{\beta N_0^2}{2}}} \right) \right] \\
&\quad \times \frac{1}{\pi\sqrt{1-W^2}}\, dW
\end{aligned} \tag{5.74}$$

where $E_b = 2\beta P_s$ denotes the average bit energy. The expressions given in (5.73) and (5.74) are thus the analytical BER for the jammed noisy DCSK system. Note that for fixed jamming power P_{jam} and noise power spectral density $N_0/2$, the BER can be reduced by making one or a combination of the following adjustments.

1. Reduce the variance of $\{x_k^2\}$ for a given P_s.
2. Increase the signal power P_s.

In the argument of each of the complimentary error functions in (5.73) and (5.74), the denominator increases with β. However, in the numerator, the variation of the terms $\beta P_{\mathrm{jam}}\cos(\pi F)$ and $\frac{P_{\mathrm{jam}}\sin(\pi F)}{\sin(\pi F/\beta)} W$ with β cannot be observed easily. Therefore, the overall effect of β on the BER performance cannot be determined simply by observing (5.73) or (5.74).

Moreover, when the jamming power is zero, i.e., $P_{\mathrm{jam}} = 0$, it can be readily shown that the bit error rate reduces to

$\mathrm{BER_{DCSK}} \mid (P_{\mathrm{jam}} = 0)$

$$= \frac{1}{2}\text{erfc}\left[\left(\frac{2\Psi}{\beta} + \frac{2N_0}{\beta P_s} + \frac{N_0^2}{2\beta P_s^2}\right)^{-\frac{1}{2}}\right] \quad (5.75)$$

$$= \frac{1}{2}\text{erfc}\left[\left(\frac{2\Psi}{\beta} + 4\left(\frac{E_b}{N_0}\right)^{-1} + 2\beta\left(\frac{E_b}{N_0}\right)^{-2}\right)^{-\frac{1}{2}}\right]. \quad (5.76)$$

where $\Psi = \text{var}[x_k^2]/P_s^2$.

Example — Consider the case where the logistic map (5.39) in Sect. 5.2.1 is used for generating the chaotic sequence. Putting (5.41) and (5.42) in (5.73) gives

BER_{DCSK}

$$= \frac{1}{4}\int_{-1}^{+1}\left[\text{erfc}\left(\frac{\frac{\beta}{2} + \beta P_{\text{jam}}\cos(\pi F) - \frac{P_{\text{jam}}\sin(\pi F)}{\sin(\pi F/\beta)}W}{\sqrt{\frac{\beta}{4} + 4\beta P_{\text{jam}}\cos^2\left(\frac{\pi F}{2}\right) + \beta N_0 + 2\beta P_{\text{jam}} N_0 + \frac{\beta N_0^2}{2}}}\right)\right.$$
$$\left. + \text{erfc}\left(\frac{\frac{\beta}{2} - \beta P_{\text{jam}}\cos(\pi F) + \frac{P_{\text{jam}}\sin(\pi F)}{\sin(\pi F/\beta)}W}{\sqrt{\frac{\beta}{4} + 4\beta P_{\text{jam}}\sin^2\left(\frac{\pi F}{2}\right) + \beta N_0 + 2\beta P_{\text{jam}} N_0 + \frac{\beta N_0^2}{2}}}\right)\right]$$
$$\times \frac{1}{\pi\sqrt{1-W^2}}\,dW. \quad (5.77)$$

5.3 Simulations and Evaluation

In this section the anti-jamming performances of the coherent CSK system and the non-coherent DCSK system are studied by computer simulations. The logistic map described in Sect. 5.2.1 has been used to generate the chaotic sequences in both cases. In particular the following aspects of performance are relevant.

- BER versus the average ratio of the bit energy to noise power spectral density (E_b/N_0) in the presence of a jamming signal;
- BER versus the ratio of the signal power to the jamming signal power (P_s/P_{jam}) for constant E_b/N_0;
- BER versus the normalized jamming frequency defined in (5.5);
- BER versus the spreading factor.

The relevant simulated BERs for the coherent CSK system are shown in Figs. 5.4 and 5.5, and those for the non-coherent DCSK system are shown in Figs. 5.6 and 5.7. In all cases, we plot also the analytical BERs obtained from the expressions derived in Sects. 5.2.1 and 5.2.2. From the figures, the consistency between the analytical BERs and the simulated BERs is clearly evidenced.

Our analyses and simulations have consistently shown that the coherent CSK system generally performs better than the non-coherent DCSK system in the presence of sinusoidal jamming signals. The following observations are worth noting.

- As expected, the BER generally decreases (improves) as the average E_b/N_0 increases and the jamming power decreases.
- For the coherent CSK system, the jamming frequency has little effect on the BER, whereas for the non-coherent DCSK system, BER experiences maxima and minima as the jamming frequency varies. In particular, the non-coherent DCSK system is most severely affected when the jamming frequency is an integer multiple of the bit frequency $(1/T_b)$. This effect can be attributed to the inherent bit structure of DCSK which splits one bit duration into two slots. If the jamming frequency is an integer multiple of the bit frequency, then for each bit, the two slots will be superposed by two identical sinusoidal signal segments (may differ by a factor "-1"), giving a high correlation of the two slots, regardless of what digital message is sent. This strongly biases the correlator output and causes great error in the recovered message.
- In general, a larger spreading factor improves the anti-jamming performance of the coherent CSK system, while no significant difference in performance is observed in the non-coherent DCSK system as the spreading factor varies. This effect can be explained as follows. In the coherent CSK system, higher the spreading factor, more accurate is the information provided by the correlator regarding the message identity, irrespective of the presence of a jamming signal. This is because the receiver is assumed to be able to regenerate a clean synchronized replica of the chaotic samples. However, for the non-coherent DCSK system, increasing the spreading factor does not necessarily improve performance because noise and/or jamming signal is equally admitted in both the reference and the information-bearing half bits.

Before closing this chapter, it is worth stressing that the study of the coherent CSK system must necessarily assume the availability of replica chaotic basis signals in the receiver. Although this assumption may not be justified in practice, the anti-jamming performance derived for the coherent CSK system in this chapter serves as benchmark data for chaos-based digital communication systems under the influence of a narrowband sinusoidal jamming signal.

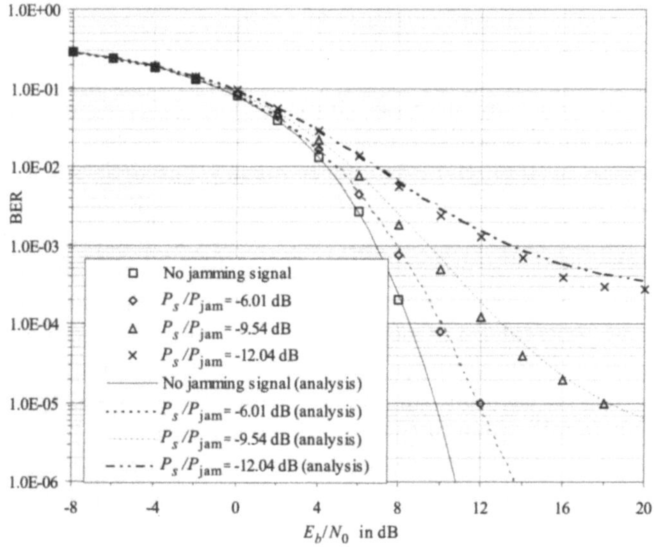

(a)

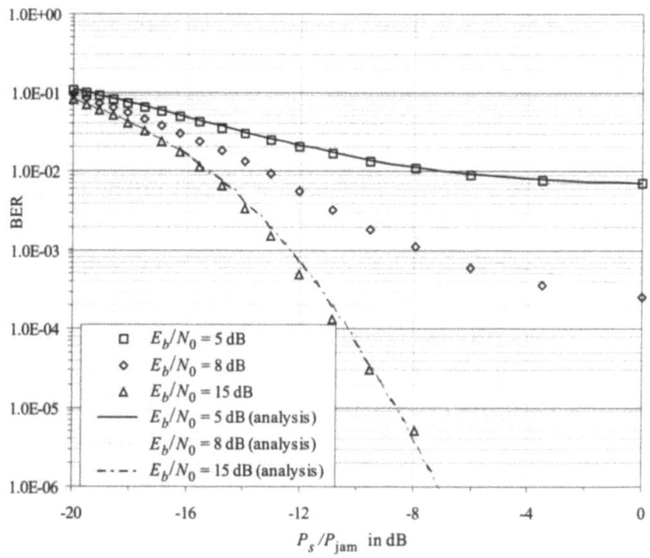

(b)

Fig. 5.4. BER of the coherent CSK system. Simulated BERs are plotted as points and analytical BERs plotted as lines. (a) BER versus E_b/N_0. Jamming level P_s/P_{jam} is -6.01, -9.54 and -12.04 dB. Spreading factor is 200. Normalized jamming frequency is 3.12735. (b) BER versus P_s/P_{jam}. Average E_b/N_0 is 5, 8 and 15 dB. Spreading factor is 200. Normalized jamming frequency is 3.12735.

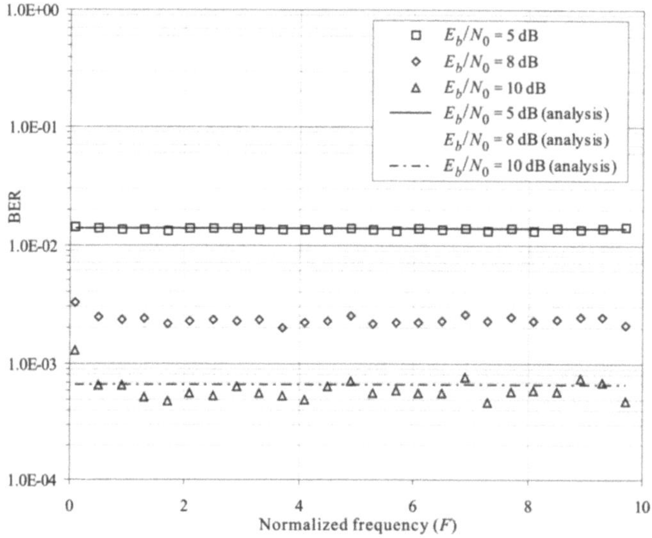

(a)

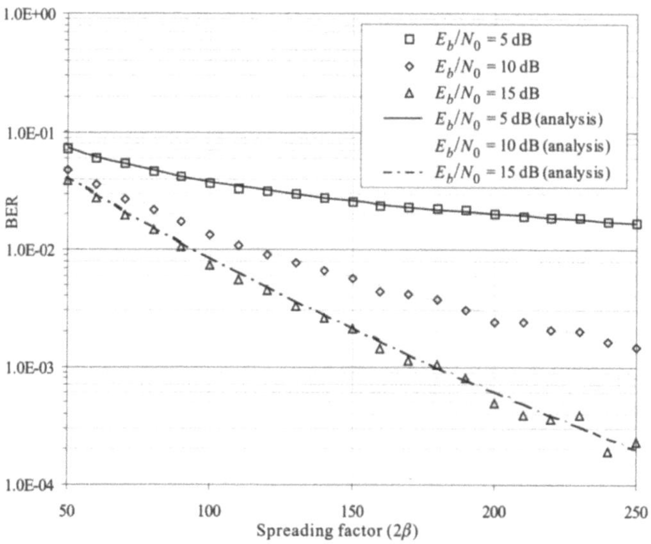

(b)

Fig. 5.5. BER of the coherent CSK system. Simulated BERs are plotted as points and analytical BERs plotted as lines. (a) BER versus the jamming frequency. Average E_b/N_0 is 5, 8 and 10 dB. Jamming level P_s/P_{jam} is -9.54 dB. Spreading factor is 200. (b) BER versus the spreading factor. Average E_b/N_0 is 5, 10 and 15 dB. Jamming level P_s/P_{jam} is -12.04 dB.

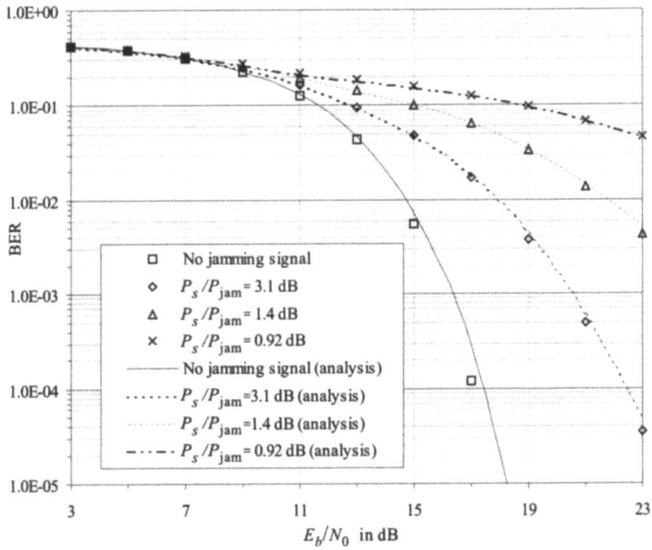

(a)

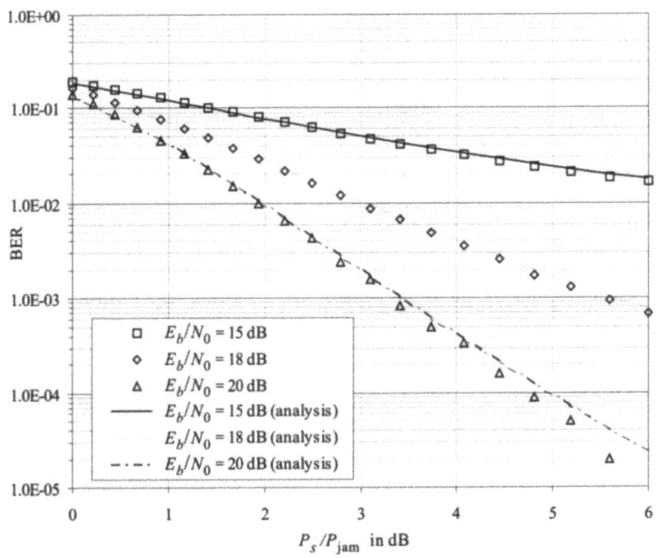

(b)

Fig. 5.6. BER of the non-coherent DCSK system. Simulated BERs are plotted as points and analytical BERs plotted as lines. (a) BER versus E_b/N_0. Jamming level P_s/P_{jam} is 0.92, 1.4 and 3.1 dB. Spreading factor is 200. Normalized jamming frequency is 3.828734. (b) BER versus P_s/P_{jam}. Average E_b/N_0 is 15, 18 and 20 dB. Spreading factor is 200. Normalized jamming frequency is 3.828734.

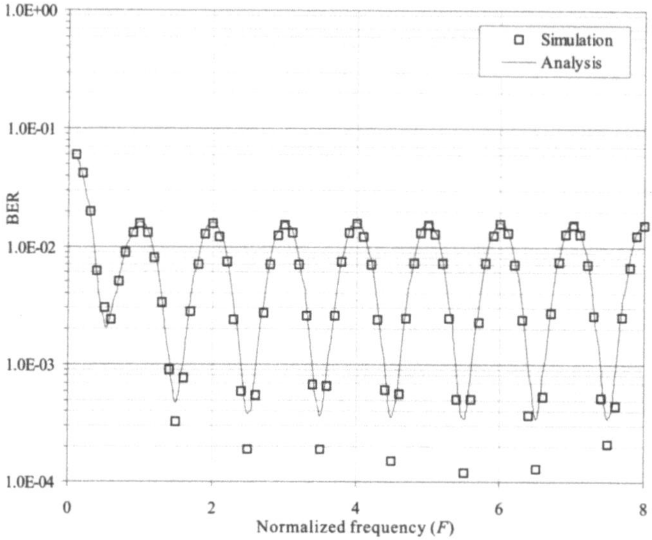

(a)

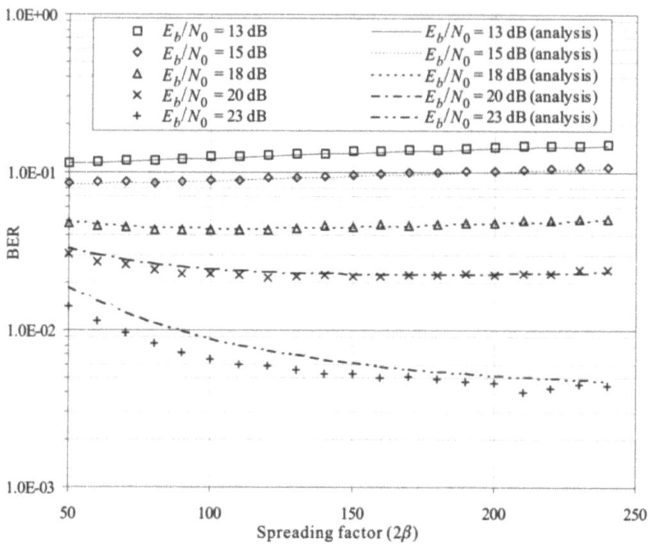

(b)

Fig. 5.7. BER of the non-coherent DCSK system. Simulated BERs are plotted as points and analytical BERs plotted as lines. (a) BER versus the normalized jamming frequency. Average E_b/N_0 is 18 dB. P_s/P_{jam} is 3.1 dB. Spreading factor is 200. (b) BER versus the spreading factor. Average E_b/N_0 is 13, 15, 18, 20 and 23 dB. Jamming level P_s/P_{jam} is 1.4 dB. Normalized jamming frequency is 3.828734.

6. Anti-Jamming Performance of Chaos-Based Digital Communication Systems Under Wideband Pulsed-Noise Jammers

In this chapter we continue our study of the anti-jamming capability of chaos-based digital communication systems. In particular, we attempt to analyze the performance of chaos-based digital communication systems under the influence of a pulsed-noise jammer, which turns on and off periodically producing a wideband jamming source [Peterson *et al.* (1995)]. The philosophy of the pulsed-noise jammer is to concentrate the jamming power during the "on" time to severely disrupt the spread-spectrum communication system. The *duty factor*, ρ, is the fraction of time during which the jammer turns on. Two types of pulsed-noise jammers are considered, namely, slowly switching and fast switching jammers. For the *slowly switching jammer*, the frequency at which the jamming signal switches between "on" and "off" is much lower than the bit frequency of the information signal. For the *fast switching jammer*, the switching frequency is assumed to be very close to the bit frequency. Within each bit period, therefore, the jammer is assumed to be turned on for a fraction ρ of the symbol period and is off for the remaining symbol period. The objective of our study is to determine the factors and the extent to which these factors would affect the performance of chaos-based systems. Analytical bit error rates (BERs) are derived, allowing the evaluation of performance for a range of noise level, jamming power, spreading factor and duty factor. For the case of slowly switching jammers, the value of ρ that maximizes the BER is also found analytically.

6.1 Systems Subject to Wideband Pulse-Noise Jammers

The basic problem we wish to investigate in this chapter is the performance of chaos-based digital communication systems when the channel is subject to a pulsed-noise jammer as well as additive white Gaussian noise. Figure 6.1 shows a block diagram of the system under study. In our analysis, we assume that the pulsed-noise jammer transmits pulses of bandlimited white Gaussian noise having an average power P_j. Also, the jamming signal occupies the same bandwidth as the chaotic signal. As mentioned before, the pulsed-noise jammer concentrates its power during the "on" time to disrupt the spread-spectrum communication system. Since the jammer turns on for a fraction ρ

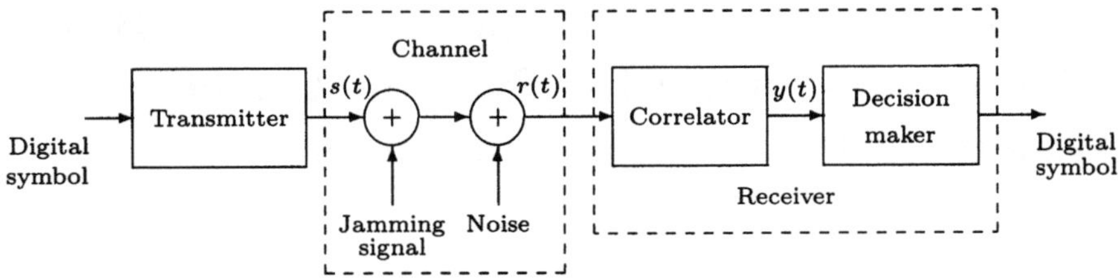

Fig. 6.1. Block diagram of a chaotic communication system with additive pulsed-noise jamming and noise.

of the time and the average jamming power is P_j, as defined earlier, the jamming power equals $\breve{P}_j = P_j/\rho$ when the jammer is turned on. Figures 6.2 (a) and (b) compare the transmitted waveforms when $\rho = 1.0$ and $\rho = 0.5$.

Suppose the transmitter output is $s(t)$. Assuming additive jamming and noise, the signal at the input of the receiver, $r(t)$, is given by

$$r(t) = s(t) + u(t) + \xi(t) \tag{6.1}$$

where $\xi(t)$ is the Gaussian noise function, and $u(t)$ represents the jamming signal which is Gaussian with zero mean and variance $\breve{P}_j$ when the jammer is turned on. The method of demodulation depends on the type of the system under study. Here, we assume that the chaotic signal can be exactly reproduced at the receiving end and thus a correlator-type receiver can be employed to decode the received signal. In this chapter we further focus on the anti-jamming performance of a coherent CSK system, similar to the one studied in Chap. 5 [Dedieu *et al.* (1993)]. Specifically, in the following section, we will attempt to develop analytical expressions for the bit error rates of the CSK system under the influence of slowly switching and fast switching pulsed-noise jammers.

6.2 Analysis of Performance Under Pulsed-Noise Jammer

We consider a binary CSK communication system, as shown in Fig. 6.3. In the transmitter, a pair of chaotic sequences, denoted by $\{\hat{x}_k\}$ and $\{\check{x}_k\}$, are generated by two chaotic maps. If the symbol "+1" is sent, $\{\hat{x}_k\}$ is transmitted during a bit period, and if "−1" is sent, $\{\check{x}_k\}$ is transmitted.

For simplicity, we consider a CSK system in which one chaos generator is used to produce chaotic signal samples $\{x_k\}$ for $k = 1, 2, \ldots$. The two possible transmitted sequences are $\{\hat{x}_k = x_k\}$ and $\{\check{x}_k = -x_k\}$. Suppose $\alpha_l \in \{-1, +1\}$ is the symbol to be sent during the lth bit period and assume that "−1" and "+1" occur with equal probabilities. Define the spreading factor, 2β, as the number of chaotic samples used to transmit one binary symbol.

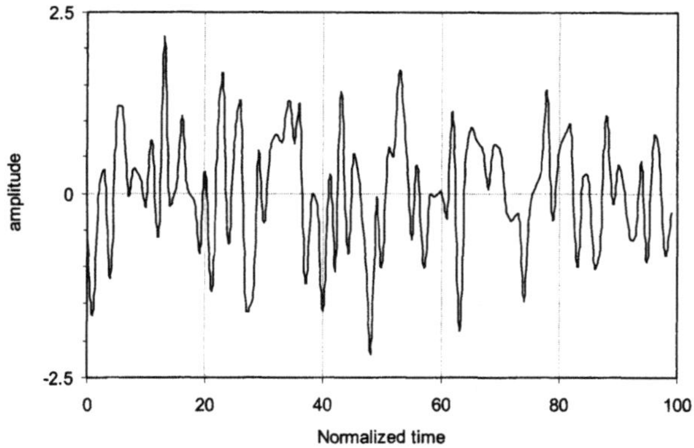

(a)

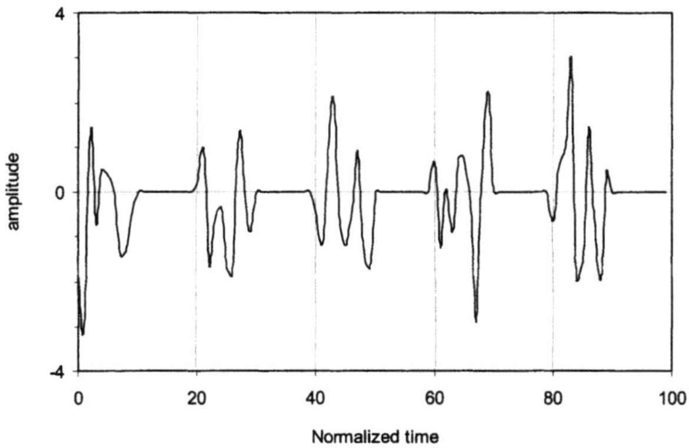

(b)

Fig. 6.2. Pulsed-noise jamming signal. (a) $\rho = 1.0$; (b) $\rho = 0.5$.

During the lth bit duration, i.e., for time $k = 2(l-1)\beta+1, 2(l-1)\beta+2, \ldots, 2l\beta$, the output of the transmitter is

$$s_k = \alpha_l x_k. \tag{6.2}$$

In addition to the white Gaussian noise, a pulsed-noise jamming signal of average power P_j and duty factor ρ is added to the transmitted signal in the channel.

Two types of pulsed-noise jammers are considered here. For the first type, the frequency at which the jamming signal turns on and off is much lower than the bit frequency of the information signal. That is to say, the period in which the jamming signal stays on or off is much longer than the bit duration [Torrieri (1986)]. We refer to this type of jammers as *slowly switching jam-*

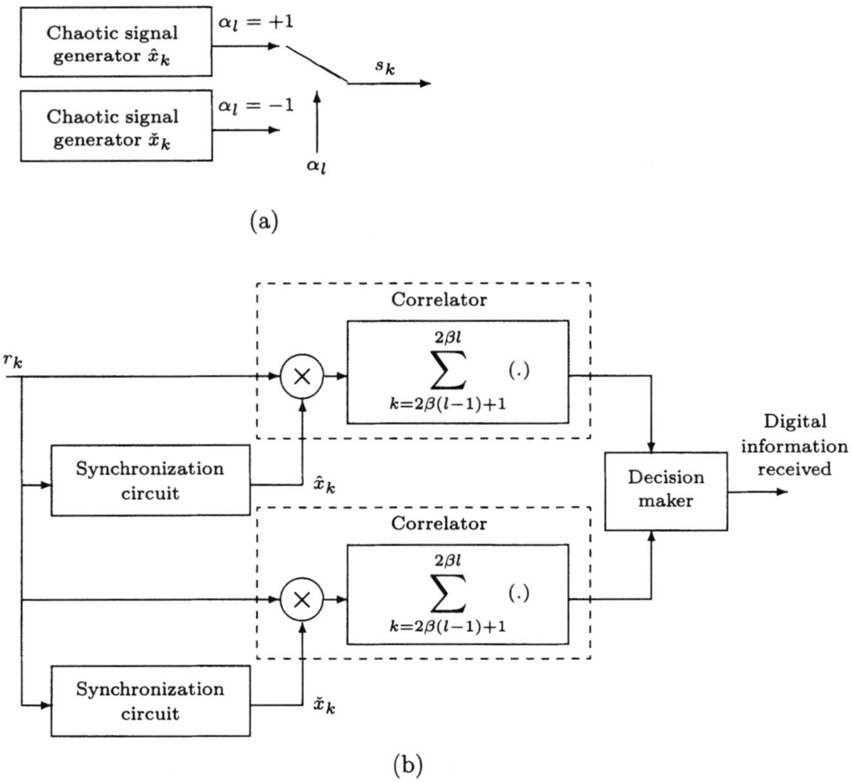

Fig. 6.3. Block diagram of a coherent CSK system. (a) Transmitter; (b) receiver.

mers. In this case, if a particular transmitted symbol is jammed, it is jammed for the whole symbol duration. The probability of a particular symbol being jammed is thus equal to ρ. For the second type of jammer, the switching frequency of the jammer is very close to the bit frequency. Within each bit period, we roughly estimate that the jammer is turned on for a fraction ρ of the symbol period and is off for the remaining symbol period. In other words, all transmitted CSK symbols are affected by the jamming signal for a fraction ρ of the symbol period. We simply refer to this type of jammers as *fast switching jammers*. The performance of the chaos-based communication system is now investigated for each of the two types of jammers.

6.2.1 Slowly Switching Jammer

Referring to Fig. 6.3, the jammed noisy signal arriving at the receiver, denoted by r_k, is given by

$$r_k = s_k + u_k + \xi_k. \tag{6.3}$$

The variable u_k represents the jamming signal, which is Gaussian with zero mean. The variance of u_k is $\check{P}_j$ when the jammer is turned on, and is zero when the jammer is off. Also, ξ_k is a Gaussian noise sample with mean equal to zero and variance (i.e., power spectral density) equal to $N_0/2$.

Consider the lth transmitted bit. Suppose the jammer is now on, i.e, the jamming power is P_j/ρ, where P_j is the average jamming power and ρ the duty factor. Thus, in (6.3), we may replace u_k by v_k, where v_k is a Gaussian random variable with zero mean and variance $\breve{P}_j = P_j/\rho$. Assuming that a correlator-type receiver is employed, the correlator output for the lth bit, y_l, is given by

$$\begin{aligned} y_l &= \sum_{k=2\beta(l-1)+1}^{2\beta l} r_k x_k \\ &= \underbrace{\alpha_l \sum_{k=2\beta(l-1)+1}^{2\beta l} x_k^2}_{\text{required signal}} + \underbrace{\sum_{k=2\beta(l-1)+1}^{2\beta l} v_k x_k}_{\text{jamming signal}} + \underbrace{\sum_{k=2\beta(l-1)+1}^{2\beta l} \xi_k x_k}_{\text{noise}} . \end{aligned} \tag{6.4}$$

Suppose a "+1" is transmitted for the lth symbol, i.e., $\alpha_l = +1$. For simplicity we write $y_l \mid (\alpha_l = +1)$ as

$$y_l \mid (\alpha_l = +1) = A + B + C \tag{6.5}$$

where A, B and C are the required signal, jamming signal and noise, respectively, and are defined as

$$A = \sum_{k=2\beta(l-1)+1}^{2\beta l} x_k^2 \tag{6.6}$$

$$B = \sum_{k=2\beta(l-1)+1}^{2\beta l} v_k x_k \tag{6.7}$$

$$C = \sum_{k=2\beta(l-1)+1}^{2\beta l} \xi_k x_k . \tag{6.8}$$

The mean value of $y_l \mid (\alpha_l = +1)$ is

$$\begin{aligned} \mathrm{E}[y_l \mid (\alpha_l = +1)] &= \mathrm{E}[A] + \mathrm{E}[B] + \mathrm{E}[C] \\ &= \mathrm{E}\left[\sum_{k=2\beta(l-1)+1}^{2\beta l} x_k^2\right] + \mathrm{E}\left[\sum_{k=2\beta(l-1)+1}^{2\beta l} v_k x_k\right] \\ &\quad + \mathrm{E}\left[\sum_{k=2\beta(l-1)+1}^{2\beta l} \xi_k x_k\right] \\ &= \sum_{k=2\beta(l-1)+1}^{2\beta l} \mathrm{E}[x_k^2] + \sum_{k=2\beta(l-1)+1}^{2\beta l} \mathrm{E}[v_k]\mathrm{E}[x_k] \end{aligned}$$

$$
\begin{aligned}
&+ \sum_{k=2\beta(l-1)+1}^{2\beta l} \mathrm{E}[\xi_k]\mathrm{E}[x_k] \\
&= 2\beta P_s
\end{aligned}
\tag{6.9}
$$

where P_s is the average power of the chaotic signal, i.e.,

$$P_s = \mathrm{E}[x_k^2]. \tag{6.10}$$

The last equality in (6.9) holds because $\mathrm{E}[v_k] = 0$ and $\mathrm{E}[\xi_k] = 0$. The variance of $y_l \mid (\alpha_l = +1)$ is

$$
\begin{aligned}
\mathrm{var}[y_l \mid (\alpha_l = +1)] &= \mathrm{var}[A] + \mathrm{var}[B] + \mathrm{var}[C] + 2\mathrm{cov}[A, B] \\
&\quad + 2\mathrm{cov}[B, C] + 2\mathrm{cov}[A, C]
\end{aligned}
\tag{6.11}
$$

where $\mathrm{cov}[X, Y]$ is the covariance of X and Y defined as

$$\mathrm{cov}[X, Y] = \mathrm{E}[XY] - \mathrm{E}[X]\mathrm{E}[Y]. \tag{6.12}$$

It can be proved that $\mathrm{cov}[A, B]$, $\mathrm{cov}[A, C]$ and $\mathrm{cov}[B, C]$ are zero (see Appendix 6A). Hence, (6.11) can be simplified to

$$\mathrm{var}[y_l \mid (\alpha_l = +1)] = \mathrm{var}[A] + \mathrm{var}[B] + \mathrm{var}[C]. \tag{6.13}$$

The average power of the chaotic signal given in (6.10) can be found by simulation, or by numerical integration if the invariant probability density function of $\{x_k\}$ is available. Also, the variance terms in (6.13) can be computed by numerical simulation. Hence, $\mathrm{E}[y_l \mid (\alpha_l = +1)]$ and $\mathrm{var}[y_l \mid (\alpha_l = +1)]$ can be evaluated.

For the lth symbol, an error occurs when $y_l \leq 0 \mid (\alpha_l = +1)$. When 2β is large, $y_l \mid (\alpha_l = +1)$ is the sum of a large number of random variables and we may assume that it follows a normal distribution. The error probability is thus given by

$$\mathrm{Prob}(y_l \leq 0 \mid (\alpha_l = +1)) = \frac{1}{2}\mathrm{erfc}\left(\frac{\mathrm{E}\left[y_l \mid (\alpha_l = +1)\right]}{\sqrt{2\mathrm{var}\left[y_l \mid (\alpha_l = +1)\right]}}\right) \tag{6.14}$$

where erfc(.) is the complementary error function defined as

$$\mathrm{erfc}(\psi) \equiv \frac{2}{\sqrt{\pi}} \int_{\psi}^{\infty} e^{-\lambda^2} d\lambda. \tag{6.15}$$

Similarly, when a "−1" is sent, the output of the correlator can be shown equal to

$$y_l \mid (\alpha_l = -1) = -A + B + C. \tag{6.16}$$

Likewise, the mean and variance of $y_l \mid (\alpha_l = -1)$, denoted by $\mathrm{E}[y_l \mid (\alpha_l = -1)]$ and $\mathrm{var}[y_l \mid (\alpha_l = -1)]$, can be derived, i.e.,

$$\mathrm{E}[y_l \mid (\alpha_l = -1)] = -2\beta P_s \tag{6.17}$$

$$\mathrm{var}[y_l \mid (\alpha_l = -1)] = \mathrm{var}[A] + \mathrm{var}[B] + \mathrm{var}[C] \tag{6.18}$$

where A, B, and C are defined in (6.6) to (6.8). The corresponding error probability is

$$\mathrm{Prob}(y_l > 0 \mid (\alpha_l = -1)) = \frac{1}{2}\mathrm{erfc}\left(\frac{-\mathrm{E}\left[y_l \mid (\alpha_l = -1)\right]}{\sqrt{2\mathrm{var}\left[y_l \mid (\alpha_l = -1)\right]}}\right). \tag{6.19}$$

Hence, the error probability of the lth transmitted symbol, given the jamming power is turned on, is

$$\begin{aligned}\mathrm{BER}_{\mathrm{on}}^{(l)} &= \mathrm{Prob}(\alpha_l = 1) \times \mathrm{Prob}(y_l \leq 0 \mid (\alpha_l = +1)) \\ &\quad + \mathrm{Prob}(\alpha_l = -1) \times \mathrm{Prob}(y_l > 0 \mid (\alpha_l = -1)) \\ &= \frac{1}{4}\mathrm{erfc}\left(\frac{\mathrm{E}\left[y_l \mid (\alpha_l = +1)\right]}{\sqrt{2\mathrm{var}\left[y_l \mid (\alpha_l = +1)\right]}}\right) \\ &\quad + \frac{1}{4}\mathrm{erfc}\left(\frac{-\mathrm{E}\left[y_l \mid (\alpha_l = -1)\right]}{\sqrt{2\mathrm{var}\left[y_l \mid (\alpha_l = -1)\right]}}\right) \\ &= \frac{1}{2}\mathrm{erfc}\left(\frac{2\beta P_s}{\sqrt{2\mathrm{var}[A] + 2\mathrm{var}[B] + 2\mathrm{var}[C]}}\right). \end{aligned} \tag{6.20}$$

Moreover, if the jamming power is off during the lth transmitted symbol period, it can be shown that the error probability is given by

$$\mathrm{BER}_{\mathrm{off}}^{(l)} = \frac{1}{2}\mathrm{erfc}\left(\frac{2\beta P_s}{\sqrt{2\mathrm{var}[A] + 2\mathrm{var}[C]}}\right) \tag{6.21}$$

which is simply (6.20) with var$[B]$ being set to zero. Since the pulsed-noise jammer has a duty factor of ρ, a transmitted symbol will be interfered by the jammer with a probability of ρ. Thus, for the lth transmitted symbol, the overall error rate is

$$\begin{aligned}\mathrm{BER}^{(l)} &= \rho \times \mathrm{BER}_{\mathrm{on}}^{(l)} + (1-\rho) \times \mathrm{BER}_{\mathrm{off}}^{(l)} \\ &= \frac{\rho}{2}\mathrm{erfc}\left(\frac{2\beta P_s}{\sqrt{2\mathrm{var}[A] + 2\mathrm{var}[B] + 2\mathrm{var}[C]}}\right) \\ &\quad + \frac{(1-\rho)}{2}\mathrm{erfc}\left(\frac{2\beta P_s}{\sqrt{2\mathrm{var}[A] + 2\mathrm{var}[C]}}\right). \end{aligned} \tag{6.22}$$

Since $\mathrm{BER}^{(l)}$ is independent of l, the bit error probability of the lth transmitted symbol is simply the bit error rate (BER) of the whole system, the latter being denoted by BER_I, i.e.,

$$\mathrm{BER}_I = \mathrm{BER}^{(l)}$$

$$= \frac{\rho}{2}\mathrm{erfc}\left(\frac{2\beta P_s}{\sqrt{2\mathrm{var}[A] + 2\mathrm{var}[B] + 2\mathrm{var}[C]}}\right) + \frac{(1-\rho)}{2}\mathrm{erfc}\left(\frac{2\beta P_s}{\sqrt{2\mathrm{var}[A] + 2\mathrm{var}[C]}}\right). \quad (6.23)$$

Hence, P_s, var$[A]$, var$[B]$ and var$[C]$ can be computed and substituted into (6.23) to obtain the BER of the system.

6.2.1.1 Analytical BERs. We begin with the assumption that the auto-variance of $\{x_k^2\}$ vanishes, i.e.,

$$\mathrm{cov}[x_k^2, x_m^2] = \mathrm{E}[x_k^2 x_m^2] - \mathrm{E}[x_k^2]\mathrm{E}[x_m^2] = 0 \quad \text{for} \quad k \neq m. \quad (6.24)$$

It then follows that the variances of the variables A, B and C are given by (see Appendix 6B)

$$\begin{aligned} \mathrm{var}[A] &= 2\beta\Lambda & (6.25)\\ \mathrm{var}[B] &= 2\beta\breve{P}_j P_s = 2\beta P_j P_s/\rho & (6.26)\\ \mathrm{var}[C] &= \beta N_0 P_s & (6.27) \end{aligned}$$

where Λ is the variance of $\{x_k^2\}$, i.e.,

$$\Lambda = \mathrm{var}[x_k^2]. \quad (6.28)$$

Putting (6.25), (6.26) and (6.27) into (6.23), the BER can be found as

$$\begin{aligned} \mathrm{BER}_I &= \frac{\rho}{2}\mathrm{erfc}\left(\frac{2\beta P_s}{\sqrt{4\beta\Lambda + 4\beta P_j P_s/\rho + 2\beta N_0 P_s}}\right) + \frac{(1-\rho)}{2}\mathrm{erfc}\left(\frac{2\beta P_s}{\sqrt{4\beta\Lambda + 2\beta N_0 P_s}}\right) \\ &= \frac{\rho}{2}\mathrm{erfc}\left[\left(\frac{\Psi}{\beta} + \frac{P_j}{\rho\beta P_s} + \frac{N_0}{2\beta P_s}\right)^{-\frac{1}{2}}\right] + \frac{(1-\rho)}{2}\mathrm{erfc}\left[\left(\frac{\Psi}{\beta} + \frac{N_0}{2\beta P_s}\right)^{-\frac{1}{2}}\right] \quad (6.29) \\ &= \frac{\rho}{2}\mathrm{erfc}\left[\left(\frac{\Psi}{\beta} + \frac{2P_j}{\rho E_b} + \frac{N_0}{E_b}\right)^{-\frac{1}{2}}\right] + \frac{(1-\rho)}{2}\mathrm{erfc}\left[\left(\frac{\Psi}{\beta} + \frac{N_0}{E_b}\right)^{-\frac{1}{2}}\right] \quad (6.30) \end{aligned}$$

where $E_b = 2\beta P_s$ denotes the average bit energy and $\Psi = \mathrm{var}[x_k^2]/P_s^2$ is a constant depending upon the type of the chaotic sequence (see Sect. 3.2.3.2 for details). The expression given in (6.29) or (6.30) is thus the analytical BER for the coherent CSK system under the influence of a slowly switching pulsed-noise jammer. Note that for fixed average jamming power P_j, duty factor ρ and noise power spectral density $N_0/2$, the BER can be improved by making one or more of the following adjustments.

1. Reduce the variance of $\{x_k^2\}$ for a given P_s.
2. Increase the spreading factor 2β.
3. Increase the bit energy E_b.

With all other parameters fixed, the second term in (6.30) decreases with ρ because of the $(1-\rho)$ factor. For the first term, increasing ρ reduces the value of the complimentary error function. However, the overall trend of the first term is uncertain because of the $\rho/2$ factor being multiplied to the complimentary error function. On the whole, the bit error rate may increase or decrease as ρ varies. When $\rho = 0$, i.e., the jamming power is zero, it can be readily shown that the bit error rate reduces to

$$\mathrm{BER}_I \mid (\rho = 0) = \frac{1}{2}\mathrm{erfc}\left[\left(\frac{\Psi}{\beta} + \frac{N_0}{E_b}\right)^{-\frac{1}{2}}\right]. \tag{6.31}$$

6.2.1.2 Maximum (Worst-Case) BERs. Suppose the CSK system is designed to operate in a jamming environment. The maximum possible average transmitter power P_s is normally used and the effect of the additive white Gaussian noise can be safely neglected. Then, the second term in (6.29) vanishes and the BER can be approximated as

$$\mathrm{BER}_I \approx \frac{\rho}{2}\mathrm{erfc}\left[\left(\frac{\Psi}{\beta} + \frac{P_j}{\rho\beta P_s} + \frac{N_0}{2\beta P_s}\right)^{-\frac{1}{2}}\right] \approx \frac{\rho}{2}\mathrm{erfc}\left[\left(\frac{\Psi}{\beta} + \frac{P_j}{\rho\beta P_s}\right)^{-\frac{1}{2}}\right]. \tag{6.32}$$

As explained in the previous section, (6.32) may increase or decrease with ρ. If we further assume that

$$\Psi \ll \frac{P_j}{\rho P_s}, \tag{6.33}$$

then (6.32) can be approximated as

$$\mathrm{BER}_I \approx \frac{\rho}{2}\mathrm{erfc}\left(\sqrt{\frac{\rho\beta P_s}{P_j}}\right). \tag{6.34}$$

Note that the complementary error function is bounded, as studied thoroughly in Wozencraft and Jacobs [1965], i.e.,

$$\mathrm{erfc}(z) \le \frac{\exp(-z^2)}{\sqrt{\pi}z}. \tag{6.35}$$

From (6.34), we have

$$\mathrm{BER}_I \le \frac{\rho}{\sqrt{4\pi\rho\beta P_s/P_j}}\exp\left(\frac{-\rho\beta P_s}{P_j}\right). \tag{6.36}$$

Table 6.1. Worst-case BERs and the corresponding duty factor $\rho_{\max}$.

Condition	Maximizing ρ	Maximum (worst-case) BER
$\frac{P_j}{2\beta P_s} \leq 1$	$\rho_{\max} = \frac{P_j}{2\beta P_s}$	$\mathrm{BER}_{\max} = \frac{1}{\sqrt{2\pi e}} \frac{P_j}{2\beta P_s}$
$\frac{P_j}{2\beta P_s} > 1$	$\rho_{\max} = 1$	$\mathrm{BER}_{\max} = \frac{1}{2}\mathrm{erfc}\left(\sqrt{\frac{\beta P_s}{P_j}}\right)$

The maximum (worst-case) value of the BER over ρ can be obtained by setting the first derivative of BER_I to zero. It can be readily shown that the value of ρ for which BER_I is maximum is given by

$$\rho_{\max} = \rho\big|_{\frac{d\mathrm{BER}_I}{d\rho}=0} = \frac{P_j}{2\beta P_s} \tag{6.37}$$

and the maximum (worst-case) BER, denoted by $\mathrm{BER}_{\max}$, is

$$\mathrm{BER}_{\max} = \frac{1}{\sqrt{2\pi e}} \frac{P_j}{2\beta P_s}. \tag{6.38}$$

Since the duty factor is always less than or equal to 1, (6.38) is valid only when $P_j/2\beta P_s \leq 1$. For $P_j/2\beta P_s > 1$, moreover, the maximum (worst-case) BER should be found from (6.34) with $\rho = 1$, i.e.,

$$\mathrm{BER}_{\max} = \begin{cases} \frac{1}{\sqrt{2\pi e}} \frac{P_j}{2\beta P_s} & \text{if } P_j/2\beta P_s \leq 1 \\ \mathrm{BER}_I|_{\rho=1} \approx \frac{1}{2}\mathrm{erfc}\left(\sqrt{\frac{\beta P_s}{P_j}}\right) & \text{if } P_j/2\beta P_s > 1. \end{cases} \tag{6.39}$$

The values of ρ for which the BER is maximum (worst) are tabulated in Table 6.1.

6.2.2 Fast Switching Jammer

In this case, the pulsed-noise jammer is turned on for approximately a fraction ρ of the symbol period and off for the remaining period. Assume that

$$\lambda = 2\rho\beta \tag{6.40}$$

where λ is an integer. Consider the lth symbol duration. The jamming signal u_k is now given by

$$u_k = \begin{cases} w_k & \text{for } k = 2(l-1)\beta+1, 2(l-1)\beta+2, \ldots, 2(l-1)\beta+\lambda \\ 0 & \text{for } k = 2(l-1)\beta+\lambda+1, \ldots, 2l\beta \end{cases} \tag{6.41}$$

where w_k is a Gaussian random variable with zero mean and variance $\breve{P}_j = P_j/\rho$. The received signal is

$$r_k = s_k + u_k + \xi_k. \tag{6.42}$$

The output of the correlator at the end of the lth symbol period is

$$\begin{aligned} y_l &= \sum_{k=2\beta(l-1)+1}^{2\beta l} r_k x_k \\ &= \underbrace{\alpha_l \sum_{k=2\beta(l-1)+1}^{2\beta l} x_k^2}_{\text{required signal}} + \underbrace{\sum_{k=2\beta(l-1)+1}^{2\beta(l-1)+\lambda} w_k x_k}_{\text{jamming signal}} + \underbrace{\sum_{k=2\beta(l-1)+1}^{2\beta l} \xi_k x_k}_{\text{noise}} . \end{aligned} \tag{6.43}$$

Suppose a "+1" is sent for the lth symbol, i.e., $\alpha_l = +1$. We rewrite (6.43) as

$$y_l \mid (\alpha_l = +1) = A + D + C \tag{6.44}$$

where A, D and C are the required signal, jamming signal and noise, respectively. Here, A and C are defined as in (6.6) and (6.8), and D is given by

$$D = \sum_{k=2\beta(l-1)+1}^{2\beta(l-1)+\lambda} w_k x_k. \tag{6.45}$$

Using a likewise procedure as in Sect. 6.2.1, it can be readily shown that the mean and variance of $y_l \mid (\alpha_l = +1)$ are

$$\mathrm{E}[y_l \mid (\alpha_l = +1)] = 2\beta P_s \tag{6.46}$$

$$\mathrm{var}[y_l \mid (\alpha_l = +1)] = \mathrm{var}[A] + \mathrm{var}[D] + \mathrm{var}[C] \tag{6.47}$$

where P_s is defined as in (6.10). When 2β is large, $y_l \mid (\alpha_l = +1)$ is the sum of a large number of random variables and can be assumed to be normal. The error probability for the lth symbol, given a "+1" is sent, is thus given by

$$\mathrm{Prob}(y_l \leq 0 \mid (\alpha_l = +1)) = \frac{1}{2}\mathrm{erfc}\left(\frac{2\beta P_s}{\sqrt{2\mathrm{var}[A] + 2\mathrm{var}[D] + 2\mathrm{var}[C]}}\right). \tag{6.48}$$

Likewise, given a "−1" is sent for the lth symbol, it can be shown that the correlator output is

$$y_l \mid (\alpha_l = -1) = -A + D + C \tag{6.49}$$

and the corresponding mean and variance are, respectively,

$$\mathrm{E}[y_l \mid (\alpha_l = -1)] = -2\beta P_s \tag{6.50}$$

and

$$\mathrm{var}[y_l \mid (\alpha_l = -1)] = \mathrm{var}[A] + \mathrm{var}[D] + \mathrm{var}[C]. \tag{6.51}$$

Assuming again that $y_l \mid (\alpha_l = -1)$ is normal, the probability that an incorrect symbol is detected is

$$\mathrm{Prob}(y_l > 0 \mid (\alpha_l = -1)) = \frac{1}{2}\mathrm{erfc}\left(\frac{2\beta P_s}{\sqrt{2\mathrm{var}[A] + 2\mathrm{var}[D] + 2\mathrm{var}[C]}}\right). \tag{6.52}$$

Thus, the overall error probability of the lth transmitted symbol, which is also the system's bit error rate, can be written as

$$\begin{aligned}\mathrm{BER}_{II} = \mathrm{BER}^{(l)} &= \mathrm{Prob}(\alpha_l = 1) \times \mathrm{Prob}(y_l \le 0 \mid (\alpha_l = +1)) \\ &\quad + \mathrm{Prob}(\alpha_l = -1) \times \mathrm{Prob}(y_l > 0 \mid (\alpha_l = -1)) \\ &= \frac{1}{2}\mathrm{erfc}\left(\frac{2\beta P_s}{\sqrt{2\mathrm{var}[A] + 2\mathrm{var}[D] + 2\mathrm{var}[C]}}\right). \end{aligned} \tag{6.53}$$

Hence, P_s, $\mathrm{var}[A]$, $\mathrm{var}[D]$ and $\mathrm{var}[C]$ can be computed and substituted into (6.53) to obtain the BER of the system.

Analytical BERs Assuming again that the autovariance of the chaotic samples $\{x_k^2\}$ vanishes, we apply (6.24) to obtain $\mathrm{var}[A] = 2\beta\Lambda$, as in (6.25). The variance of C is given in (6.27) whereas that of D can be shown equal to

$$\mathrm{var}[D] = \lambda \breve{P}_j P_s = 2\beta P_j P_s. \tag{6.54}$$

Substituting (6.25), (6.27) and (6.54) into (6.53), the analytical BER can be found as

$$\begin{aligned}\mathrm{BER}_{II} &= \frac{1}{2}\mathrm{erfc}\left(\frac{2\beta P_s}{\sqrt{4\beta\Lambda + 4\beta P_j P_s + 2\beta N_0 P_s}}\right) \\ &= \frac{1}{2}\mathrm{erfc}\left[\left(\frac{\Psi}{\beta} + \frac{P_j}{\beta P_s} + \frac{N_0}{2\beta P_s}\right)^{-\frac{1}{2}}\right] \qquad (6.55) \\ &= \frac{1}{2}\mathrm{erfc}\left[\left(\frac{\Psi}{\beta} + \frac{2P_j}{E_b} + \frac{N_0}{E_b}\right)^{-\frac{1}{2}}\right] \qquad (6.56)\end{aligned}$$

and $E_b = 2\beta P_s$ denotes the average bit energy. From (6.55) and (6.56), it can be observed that the BER of the system is independent of the duty factor ρ. For a given average jamming power and a given noise power spectral density, the BER can be improved by making the same adjustments suggested in Sect. 6.2.1.1.

6.3 Simulations and Evaluation

In this section the performance of the coherent CSK system under the influence of a pulsed-noise jammer is studied by computer simulations. The Chebyshev map of degree $\mu > 1$, defined as [Kohda and Tsuneda (1994)]

$$g_\mu(x) = \cos(\mu \cos^{-1}(x)) \qquad -1 \le x \le 1, \tag{6.57}$$

has been used to generate the chaotic sequences. Using this category of maps, it can be proved that the autovariance of the chaotic samples $\{x_k^2\}$ vanishes (see Appendix 6C) and hence the condition assumed in (6.24) is justified. Also, the signal power can be shown equal to

$$P_s = \mathrm{E}[x_k^2] = 0.5 \tag{6.58}$$

and the variance of $\{x_k^2\}$ is

$$\Lambda = \mathrm{var}[x_k^2] = 0.125. \tag{6.59}$$

In particular, we use the Chebyshev map of degree 3, i.e., $\mu = 3$, in our simulations and the following aspects of performance are studied in the CSK system under the influence of slowly switching and fast switching jamming signals.

- BERs versus the ratio of the average jamming signal power to signal power (P_j/P_s) for different duty factors ρ under constant E_b/N_0;
- BERs versus E_b/N_0 for different P_j/P_s under constant ρ.

For the case where the jammer is slowly switching, the theoretical maximum (worst-case) BERs are also compared with the ones obtained from simulations.

6.3.1 Slowly Switching Jammer

Figures 6.4 and 6.5 show the BERs versus P_j/P_s for different duty factors ρ. Here, we fix E_b/N_0 at 20 dB. Analytical BERs are obtained by substituting (6.58) and (6.59) into (6.29). Note that BERs less than 10^{-5} have been omitted for clarity, and therefore different horizontal scales have been used in the figures. From Figs. 6.4 and 6.5, we observe that the BERs generally increase as P_j/P_s increases (higher jamming power). Also, the BERs improve (decrease) when the spreading factor increases. These observations are consistent with our analysis in Sect. 6.2.1.1. Furthermore, the analytical BERs agree very well with the simulation results for large spreading factors (20 and 50). This is because the validity of the assumption of normal distribution of the conditional correlator output holds better for higher spreading factors.

In Figs. 6.6 and 6.7, the BERs versus E_b/N_0 for different P_j/P_s are shown. The duty factor ρ is fixed at 0.5. When P_j/P_s is small (e.g., 0 or 10 dB), the

BERs improve as E_b/N_0 increases. When P_j/P_s is large (e.g., 20 or 30 dB), the BERs stay more or less the same irrespective of the variation of E_b/N_0. This is apparently due to the dominating effect of the jamming signal on the BERs. Furthermore, we observe that the analytical BERs are in very good agreement with the simulated ones, especially for large spreading factors.

Figures 6.8 and 6.9 show the simulated BERs versus P_j/P_s for different duty factors. The maximum (worst-case) BERs tabulated in Table 6.1 are also included in the figures for comparison. We observe that the value of ρ that maximizes the BERs varies as P_j/P_s changes. In almost all cases, the simulated BERs are bounded above by the worst-case values of the BERs, except in Fig. 6.8 (a) where the simulated BERs exceed the worst-case values for P_j/P_s below -22 dB. However, for these violating cases, the assumption stated in (6.33), i.e., $\Psi \ll P_j/\rho P_s$ no longer holds, thereby invalidating the theoretical maximum (worst-case) BERs.

6.3.2 Fast Switching Jammer

In Figs. 6.10 and 6.11, the simulated BERs versus P_j/P_s for different ρ are plotted. The analytical BERs, computed by substituting (6.58) and (6.59) into (6.55), are included for comparison. From these figures, we observe that the simulated BERs are independent of the duty factor ρ. For small spreading factors ($2\beta = 5$ and 10), the usual assumption about the conditional correlator output does not hold, causing large discrepancy between the simulated and analytical BERs. However, for large spreading factors ($2\beta = 20$ and 50), the simulated and analytical BERs agree.

Figures 6.12 and 6.13 show the BERs versus E_b/N_0 for different P_j/P_s. The duty factor ρ is fixed at 1.0. Similar to the case where the jammer is slowly switching, when P_j/P_s is small (e.g., 0 or 10 dB), the BERs improve as E_b/N_0 increases. When P_j/P_s is large (e.g., 20 or 30 dB), little change is observed on the BERs as E_b/N_0 increases. We also observe that the analytical BERs generally agree with the simulated ones, especially for large spreading factors.

It should be stressed that although the anti-jamming performance may vary from system to system depending upon the exact type of the chaos generators used, the general conclusions regarding the effects of spreading factor, jamming power, signal power, etc. should remain universally valid. Further, it has been observed that the discrepancy between the simulation results and the theoretical BERs can be quite large when the spreading factor is small. This is because the assumption on the normally distributed conditional correlator output does not hold well for small spreading factors. Although exact expressions of BERs may be obtained, for example using the method of Lawrance and Ohama [2002], the computational effort for calculating BERs using such exact expressions is usually quite high.

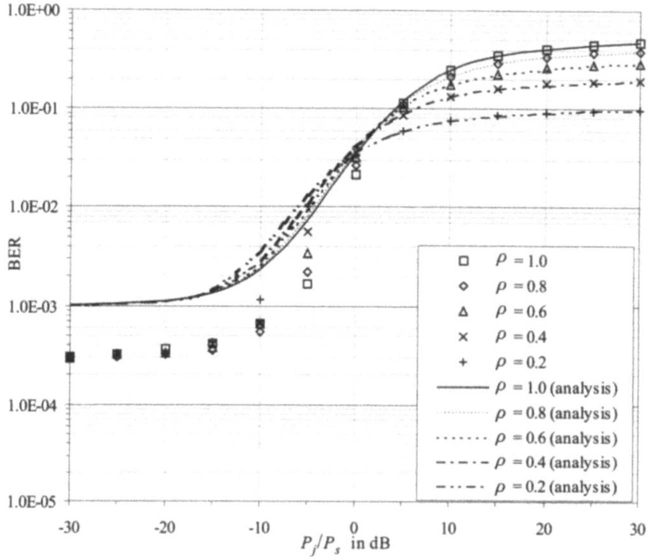

(a)

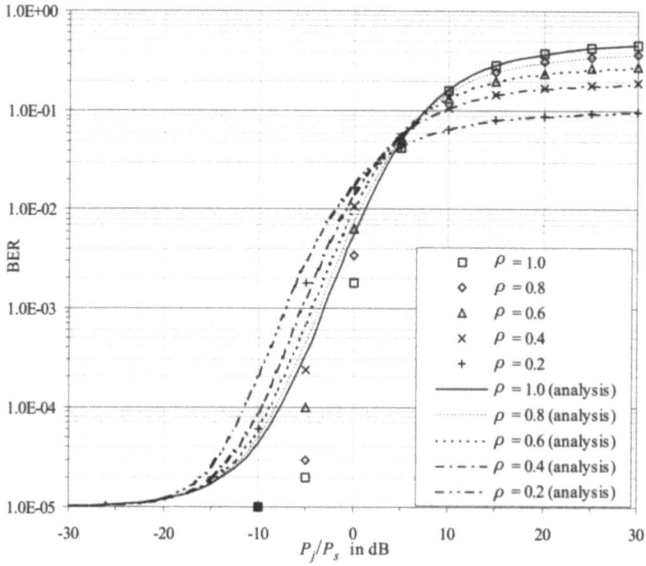

(b)

Fig. 6.4. BERs versus P_j/P_s of the coherent CSK system with a slowly switching pulsed-noise jammer. Simulated BERs are plotted as points and analytical BERs plotted as lines. $E_b/N_0 = 20$ dB and $\rho = 0.2$, 0.4, 0.6, 0.8 and 1.0. (a) Spreading factor is 5; (b) spreading factor is 10.

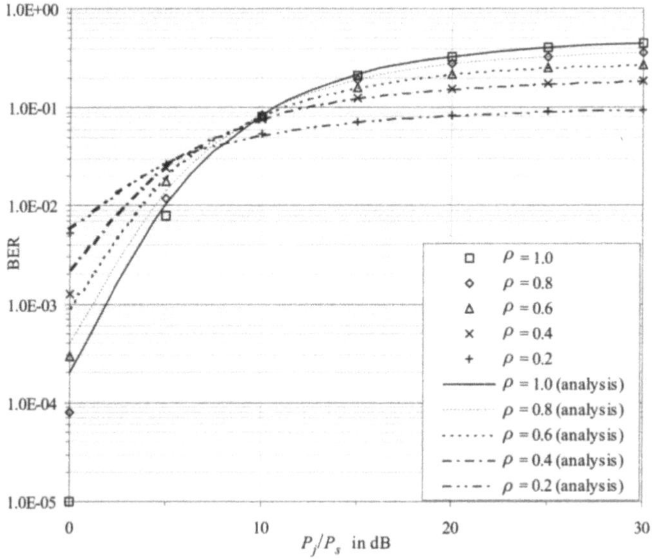

(a)

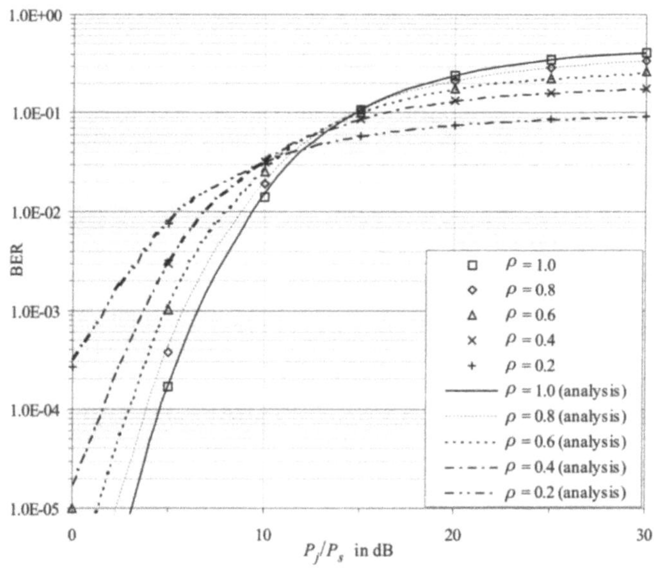

(b)

Fig. 6.5. BERs versus P_j/P_s of the coherent CSK system with a slowly switching pulsed-noise jammer. Simulated BERs are plotted as points and analytical BERs plotted as lines. $E_b/N_0 = 20$ dB and $\rho = 0.2$, 0.4, 0.6, 0.8 and 1.0. (a) Spreading factor is 20; (b) spreading factor is 50.

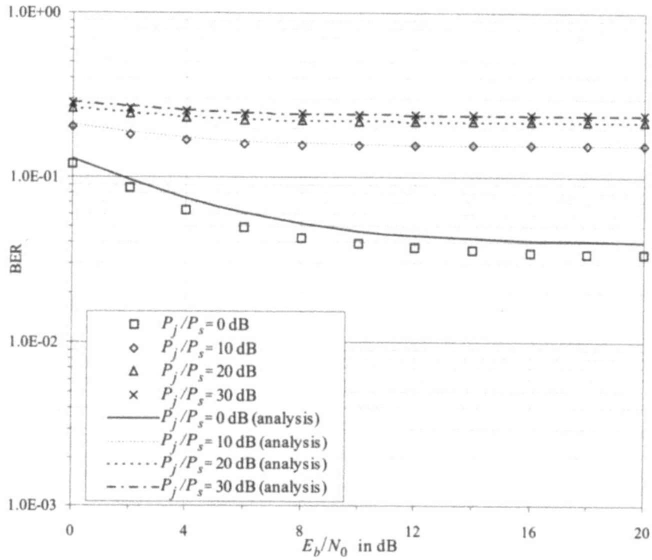

(a)

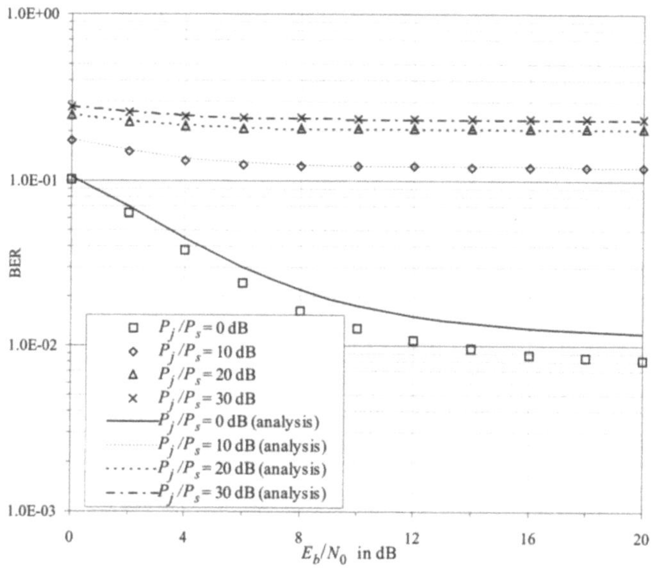

(b)

Fig. 6.6. BERs versus E_b/N_0 of the coherent CSK system with slowly switching pulsed-noise jammer. Simulated BERs are plotted as points and analytical BERs plotted as lines. $\rho = 0.5$ and $P_j/P_s = 0$, 10, 20 and 30 dB. (a) Spreading factor is 5; (b) spreading factor is 10.

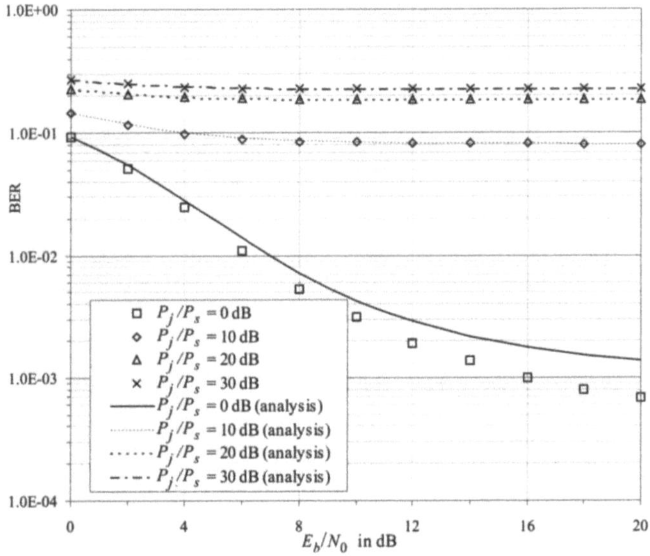

(a)

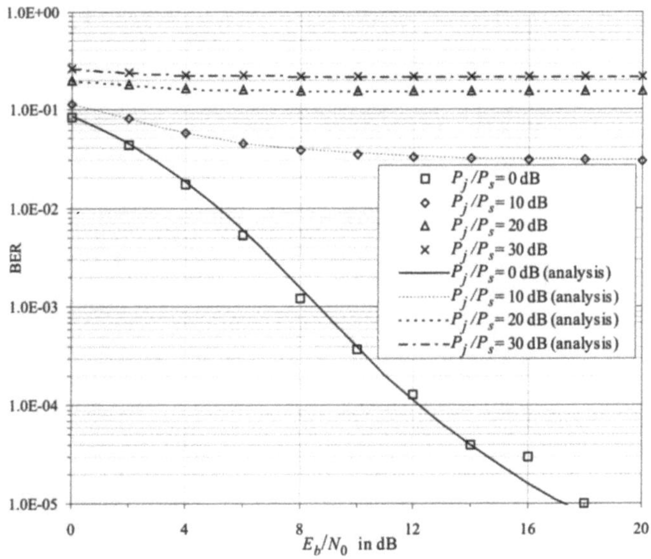

(b)

Fig. 6.7. BERs versus E_b/N_0 of the coherent CSK system with slowly switching pulsed-noise jammer. Simulated BERs are plotted as points and analytical BERs plotted as lines. $\rho = 0.5$ and $P_j/P_s = 0$, 10, 20 and 30 dB. (a) Spreading factor is 20; (b) spreading factor is 50.

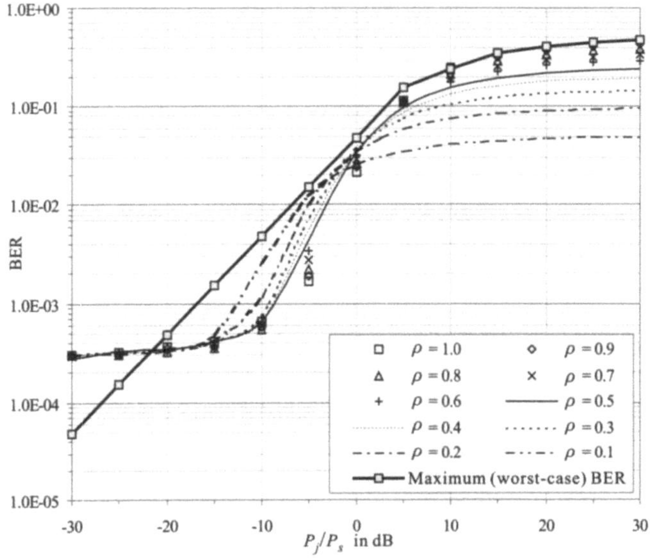

(a)

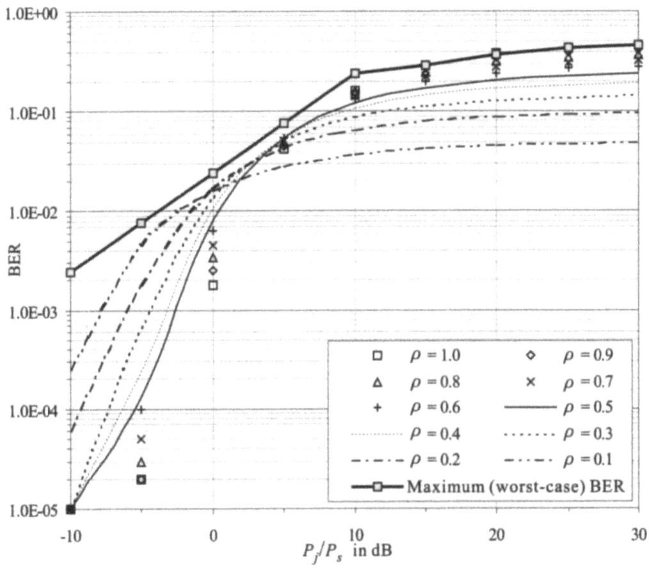

(b)

Fig. 6.8. Simulated BERs versus P_j/P_s of the coherent CSK system with a slowly switching pulsed-noise jammer. $E_b/N_0 = 20$ dB and $\rho = 0.1$ to 1.0 in steps of 0.1. (a) Spreading factor is 5; (b) spreading factor is 10.

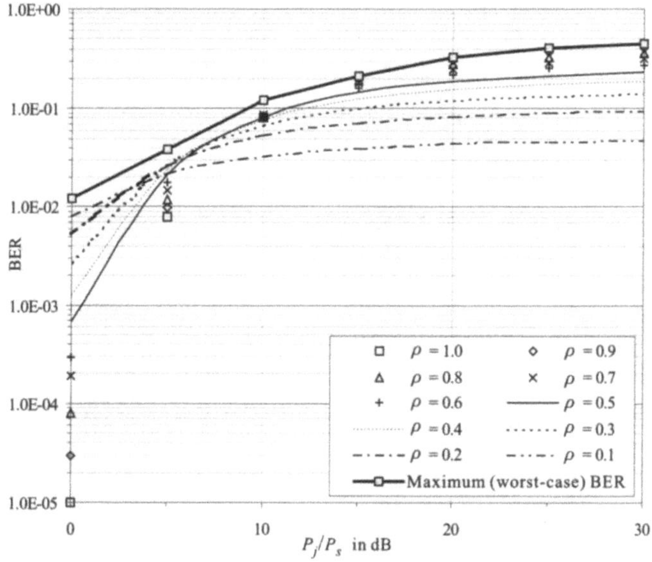

(a)

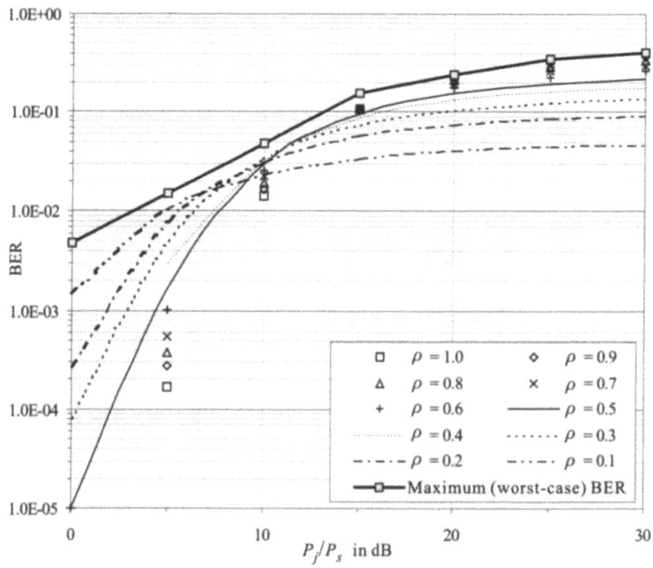

(b)

Fig. 6.9. Simulated BERs versus P_j/P_s of the coherent CSK system with a slowly switching pulsed-noise jammer. $E_b/N_0 = 20$ dB and $\rho = 0.1$ to 1.0 in steps of 0.1. (a) Spreading factor is 20; (b) spreading factor is 50.

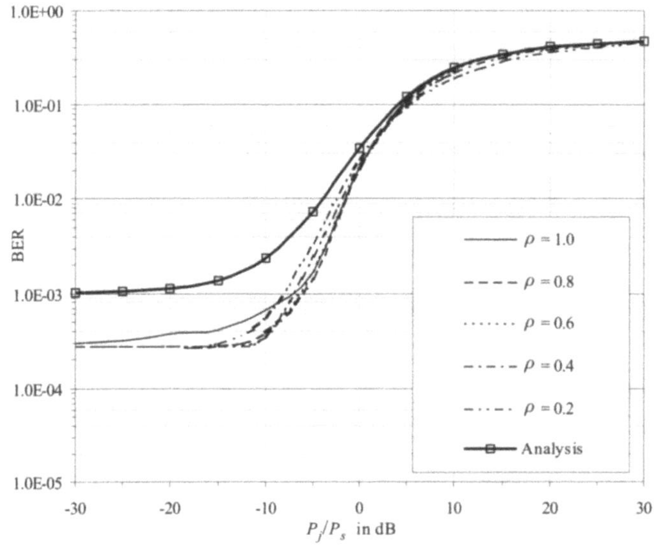

(a)

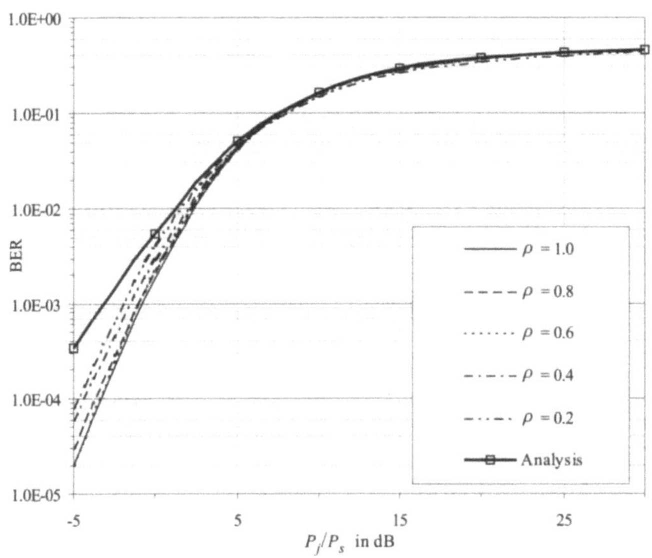

(b)

Fig. 6.10. BERs versus P_j/P_s of the coherent CSK system with a fast switching pulsed-noise jammer. $E_b/N_0 = 20$ dB and $\rho = 0.2$, 0.4, 0.6, 0.8 and 1.0. Analytical BER is independent of ρ. (a) Spreading factor is 5; (b) spreading factor is 10.

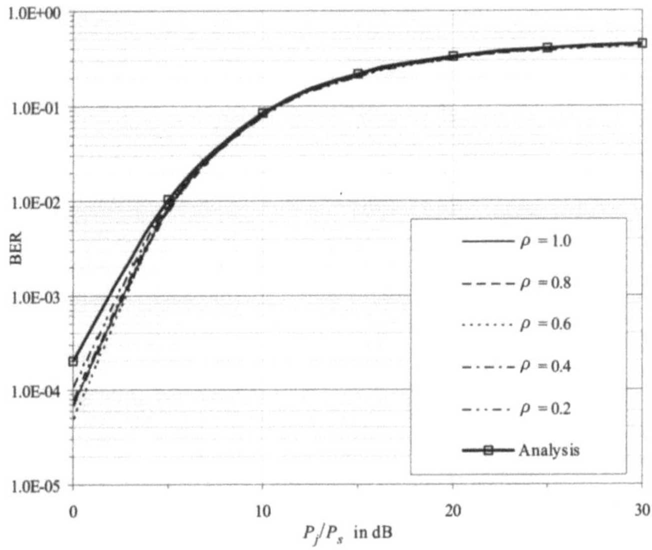

(a)

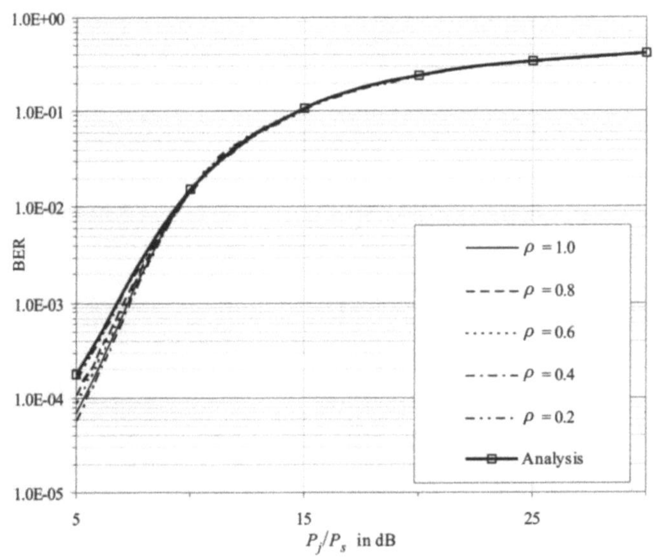

(b)

Fig. 6.11. BERs versus P_j/P_s of the coherent CSK system with a fast switching pulsed-noise jammer. $E_b/N_0 = 20$ dB and $\rho = 0.2$, 0.4, 0.6, 0.8 and 1.0. Analytical BER is independent of ρ. (a) Spreading factor is 20; (b) spreading factor is 50.

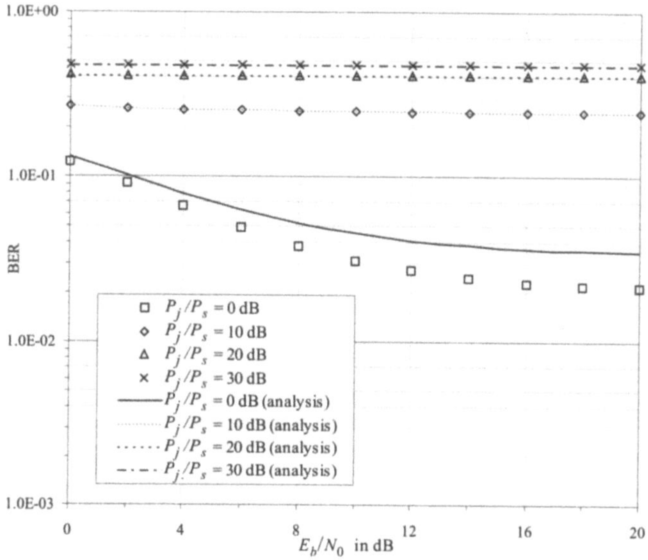

(a)

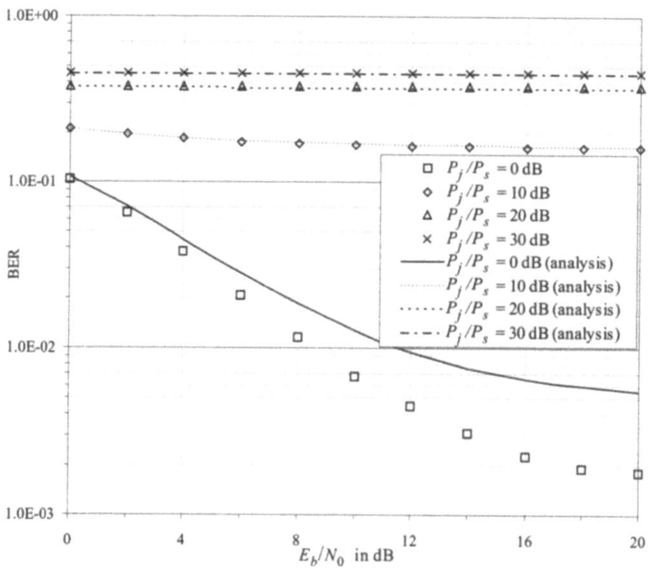

(b)

Fig. 6.12. BERs versus E_b/N_0 of the coherent CSK system with a fast switching pulsed-noise jammer. Simulated BERs are plotted as points and analytical BERs plotted as lines. $\rho = 1.0$ and $P_j/P_s = 0$, 10, 20 and 30 dB. (a) Spreading factor is 5; (b) spreading factor is 10.

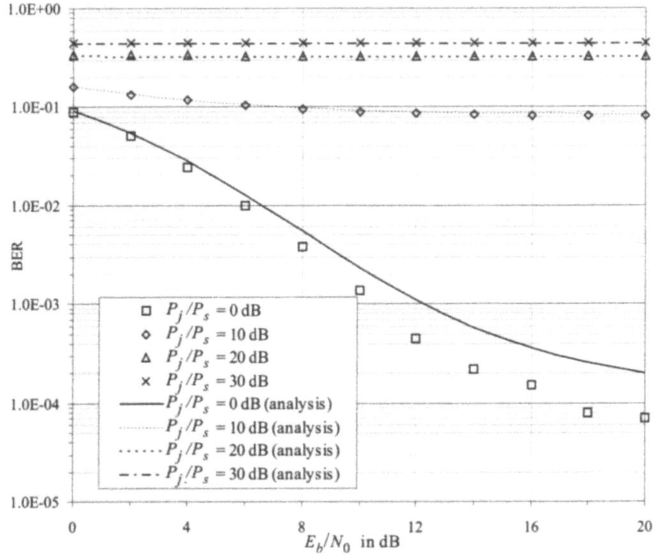

(a)

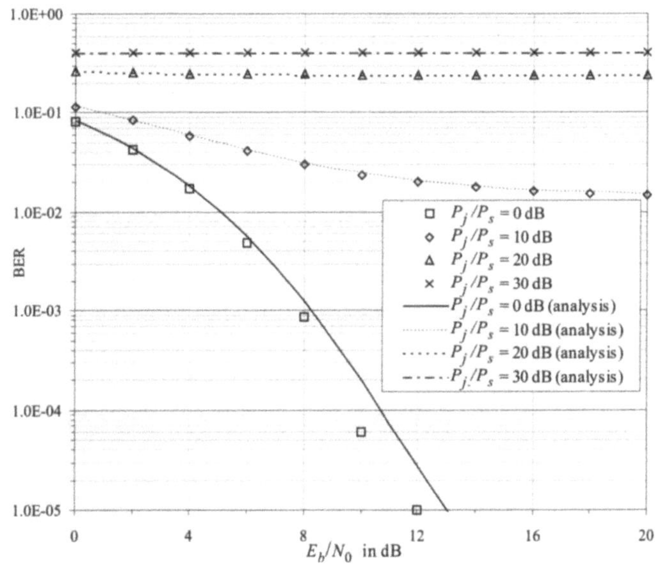

(b)

Fig. 6.13. BERs versus E_b/N_0 of the coherent CSK system with a fast switching pulsed-noise jammer. Simulated BERs are plotted as points and analytical BERs plotted as lines. $\rho = 1.0$ and $P_j/P_s = 0$, 10, 20 and 30 dB. (a) Spreading factor is 20; (b) spreading factor is 50.

Appendix 6A: Derivation of covariances

All symbols are defined as in Sect. 6.2.1.

$$\begin{aligned}
\mathrm{cov}[A,B] &= \mathrm{E}[AB]-\mathrm{E}[A]\mathrm{E}[B] \\
&= \mathrm{E}\left[\left(\sum_{k=2\beta(l-1)+1}^{2\beta l} x_k^2\right)\left(\sum_{k=2\beta(l-1)+1}^{2\beta l} v_k x_k\right)\right] \\
&\quad - \mathrm{E}\left[\sum_{k=2\beta(l-1)+1}^{2\beta l} x_k^2\right]\mathrm{E}\left[\sum_{k=2\beta(l-1)+1}^{2\beta l} v_k x_k\right] \\
&= \mathrm{E}\left[\sum_{k=2\beta(l-1)+1}^{2\beta l} x_k^3 v_k + \sum_{\substack{k=2\beta(l-1)+1\\k\neq m}}^{2\beta l}\sum_{m=2\beta(l-1)+1}^{2\beta l} v_k x_k x_m^2\right] \\
&\quad - \mathrm{E}\left[\sum_{k=2\beta(l-1)+1}^{2\beta l} x_k^2\right]\left(\sum_{k=2\beta(l-1)+1}^{2\beta l} \mathrm{E}[v_k]\mathrm{E}[x_k]\right) \\
&= \sum_{k=2\beta(l-1)+1}^{2\beta l} \mathrm{E}[x_k^3]\mathrm{E}[v_k] \\
&\quad + \sum_{\substack{k=2\beta(l-1)+1\\k\neq m}}^{2\beta l}\sum_{m=2\beta(l-1)+1}^{2\beta l} \mathrm{E}[v_k]\mathrm{E}[x_k x_m^2] \\
&= 0
\end{aligned} \tag{6.60}$$

Likewise, it can be proved that $\mathrm{cov}[A,C]=0$.

$$\begin{aligned}
\mathrm{cov}[B,C] &= \mathrm{E}[BC]-\mathrm{E}[B]\mathrm{E}[C] \\
&= \mathrm{E}\left[\left(\sum_{k=2\beta(l-1)+1}^{2\beta l} v_k x_k\right)\left(\sum_{k=2\beta(l-1)+1}^{2\beta l} \xi_k x_k\right)\right] \\
&\quad - \mathrm{E}\left[\sum_{k=2\beta(l-1)+1}^{2\beta l} v_k x_k\right]\mathrm{E}\left[\sum_{k=2\beta(l-1)+1}^{2\beta l} \xi_k x_k\right] \\
&= \mathrm{E}\left[\sum_{k=2\beta(l-1)+1}^{2\beta l} x_k^2 v_k \xi_k\right. \\
&\quad \left. + \sum_{\substack{k=2\beta(l-1)+1\\k\neq m}}^{2\beta l}\sum_{m=2\beta(l-1)+1}^{2\beta l} v_k x_k \xi_m x_m\right]
\end{aligned}$$

$$
\begin{aligned}
&- \left(\sum_{k=2\beta(l-1)+1}^{2\beta l} \mathrm{E}[v_k]\mathrm{E}[x_k] \right) \left(\sum_{k=2\beta(l-1)+1}^{2\beta l} \mathrm{E}[\xi_k]\mathrm{E}[x_k] \right) \\
&= \sum_{k=2\beta(l-1)+1}^{2\beta l} \mathrm{E}[x_k^2]\mathrm{E}[v_k]\mathrm{E}[\xi_k] \\
&\quad + \sum_{\substack{k=2\beta(l-1)+1 \\ k\neq m}}^{2\beta l} \sum_{m=2\beta(l-1)+1}^{2\beta l} \mathrm{E}[v_k]\mathrm{E}[x_k x_m]\mathrm{E}[\xi_m] \\
&= 0 \qquad (6.61)
\end{aligned}
$$

Appendix 6B: Derivation of variances

All symbols are defined as in Sect. 6.2.1.

$$
\begin{aligned}
\mathrm{var}[A] &= \mathrm{var}\left[\sum_{k=2\beta(l-1)+1}^{2\beta l} x_k^2 \right] \\
&= \sum_{k=2\beta(l-1)+1}^{2\beta l} \mathrm{var}[x_k^2] + \sum_{\substack{k=2\beta(l-1)+1 \\ k\neq m}}^{2\beta l} \sum_{m=2\beta(l-1)+1}^{2\beta l} \mathrm{cov}[x_k^2, x_m^2] \\
&= 2\beta \mathrm{var}[x_k^2] \qquad \text{apply (6.24)} \\
&= 2\beta \Lambda \qquad \text{where } \Lambda = \mathrm{var}[x_k^2] \qquad (6.62)
\end{aligned}
$$

$$
\begin{aligned}
\mathrm{var}[B] &= \mathrm{var}\left[\sum_{k=2\beta(l-1)+1}^{2\beta l} v_k x_k \right] \\
&= \sum_{k=2\beta(l-1)+1}^{2\beta l} \mathrm{var}[v_k x_k] \\
&\quad + \sum_{\substack{k=2\beta(l-1)+1 \\ k\neq m}}^{2\beta l} \sum_{m=2\beta(l-1)+1}^{2\beta l} \mathrm{cov}[v_k x_k, v_m x_m] \\
&= \sum_{k=2\beta(l-1)+1}^{2\beta l} \left\{ \mathrm{E}[v_k^2 x_k^2] - \mathrm{E}^2[v_k x_k] \right\} \\
&\quad + \sum_{\substack{k=2\beta(l-1)+1 \\ k\neq m}}^{2\beta l} \sum_{m=2\beta(l-1)+1}^{2\beta l} \{ \mathrm{E}[v_k x_k v_m x_m] \\
&\qquad - \mathrm{E}[v_k x_k]\mathrm{E}[v_m x_m] \} \\
&= \sum_{k=2\beta(l-1)+1}^{2\beta l} \left\{ \mathrm{E}[v_k^2]\mathrm{E}[x_k^2] - \mathrm{E}^2[v_k]\mathrm{E}^2[x_k] \right\}
\end{aligned}
$$

$$
\begin{aligned}
&+ \sum_{\substack{k=2\beta(l-1)+1 \\ k\neq m}}^{2\beta l} \sum_{m=2\beta(l-1)+1}^{2\beta l} \{\mathrm{E}[x_k x_m]\mathrm{E}[v_k v_m] \\
&\qquad\qquad - \mathrm{E}[x_k]\mathrm{E}[x_m]\mathrm{E}[v_k]\mathrm{E}[v_m]\} \\
&= 2\beta \breve{P}_j P_s \qquad (6.63)
\end{aligned}
$$

Likewise, it can be proved that $\mathrm{var}[C] = \beta N_0 P_s$.

Appendix 6C: Derivation of the statistical properties for the chaotic sequences generated by Chebyshev maps of degree larger than one

In this appendix, we derive the statistical properties for the chaotic sequences generated by Chebyshev maps. All symbols are defined as in Sect. 6.3. A Chebyshev map of degree μ is defined as [Kohda and Tsuneda (1994)]

$$x_{k+1} = h(x_k) = \cos(\mu \cos^{-1} x_k) \tag{6.64}$$

where μ is an integer. We consider the case where $\mu > 1$. The invariant probability density function of $\{x_k\}$, denoted by $\rho(x)$, is known to be [Kohda and Tsuneda (1994)]

$$\rho(x) = \begin{cases} \dfrac{1}{\pi\sqrt{1-x^2}} & \text{if } |x| < 1 \\ 0 & \text{otherwise.} \end{cases} \tag{6.65}$$

Derivation of $\mathrm{E}[x_k]$

Since $\rho(x)$ is an even function, the mean value of $\{x_k\}$ is

$$\mathrm{E}[x_k] = \int_{-\infty}^{\infty} x\rho(x)dx = \int_{-1}^{1} x\rho(x)dx = 0. \tag{6.66}$$

Derivation of $\mathrm{cov}[x_j, x_k]$

The autovariance of $\{x_k\}$ is given by

$$\mathrm{cov}[x_j, x_k] = \mathrm{E}[x_j x_k] - \mathrm{E}[x_j]\mathrm{E}[x_k] = \mathrm{E}[x_j x_k]. \tag{6.67}$$

We consider the case where $j \neq k$. Without loss of generality, assume $k = j+n$ for some positive integer n. Define

$$
\begin{aligned}
h^{(1)}(x) &= h(x) \\
h^{(2)}(x) &= h(h^{(1)}(x)) \\
\vdots &= \vdots
\end{aligned}
$$

$$h^{(n)}(x) \quad = \quad h(h^{(n-1)}(x)). \tag{6.68}$$

The autovariance of $\{x_k\}$ can be re-written as

$$\begin{aligned} \operatorname{cov}[x_j, x_k] &= \int_{-\infty}^{\infty} \rho(x)\ x\ h^{(n)}(x)dx \\ &= \int_{-1}^{1} \frac{1}{\pi\sqrt{1-x^2}}\ x\ h^{(n)}(x)dx. \end{aligned} \tag{6.69}$$

Making the substitution $x = \cos\phi$, (6.69) becomes

$$\begin{aligned} \operatorname{cov}[x_j, x_k] &= \int_{\pi}^{0} \frac{1}{\pi \sin\phi} \cos\phi\ h^{(n)}(\cos\phi)\ (-\sin\phi)\ d\phi \\ &= \frac{1}{\pi}\int_{0}^{\pi} \cos\phi\ h^{(n)}(\cos\phi)\ d\phi \\ &= \frac{1}{\pi}\int_{0}^{\pi} \cos\phi\ \cos(\mu^n\phi)\ d\phi \\ &= \frac{1}{2\pi}\int_{0}^{\pi} \cos((\mu^n+1)\phi) + \cos((\mu^n-1)\phi)\ d\phi \\ &= \frac{1}{2\pi}\left[\frac{\sin((\mu^n+1)\phi)}{\mu^n+1} + \frac{\sin((\mu^n-1)\phi)}{\mu^n-1}\right]_0^{\pi} \\ &= 0. \end{aligned} \tag{6.70}$$

Derivation of $\operatorname{cov}[x_j^2, x_k^2]$

The autovariance of $\{x_k^2\}$ is given by

$$\operatorname{cov}[x_j^2, x_k^2] = \mathrm{E}[x_j^2 x_k^2] - \mathrm{E}[x_j^2]\mathrm{E}[x_k^2]. \tag{6.71}$$

We consider the case where $j \neq k$. Without loss of generality, assume $k = j+n$ for some positive integer n.

$$\begin{aligned} \mathrm{E}[x_j^2 x_k^2] &= \int_{-\infty}^{\infty} \rho(x)\ x^2\ (h^{(n)}(x))^2 dx \\ &= \int_{-1}^{1} \frac{1}{\pi\sqrt{1-x^2}}\ x^2\ (h^{(n)}(x))^2 dx. \end{aligned} \tag{6.72}$$

Making the substitution $x = \cos\phi$, (6.72) becomes

$$\begin{aligned} \mathrm{E}[x_j^2 x_k^2] &= \int_{\pi}^{0} \frac{1}{\pi \sin\phi} \cos^2\phi\ \ (h^{(n)}(\cos\phi))^2\ (-\sin\phi)\ d\phi \\ &= \frac{1}{\pi}\int_{0}^{\pi} \cos^2\phi\ \ (h^{(n)}(\cos\phi))^2\ d\phi \\ &= \frac{1}{\pi}\int_{0}^{\pi} \cos^2\phi\ \cos^2(\mu^n\phi)\ d\phi \\ &= \frac{1}{\pi}\int_{0}^{\pi} \left(\frac{1+\cos 2\phi}{2}\right)\left(\frac{1+\cos(2\mu^n\phi)}{2}\right)\ d\phi \end{aligned}$$

$$
\begin{aligned}
&= \frac{1}{4\pi}\int_0^{\pi}\left(1+\cos 2\phi+\cos(2\mu^n\phi)+\frac{1}{2}\cos(2\phi(\mu^n+1))\right.\\
&\qquad\left.+\frac{1}{2}\cos(2\phi(\mu^n-1))\right)\,d\phi\\
&= \frac{1}{4\pi}\left[\phi+\frac{\sin 2\phi}{2}+\frac{\sin(2\mu^n\phi)}{2\mu^n}+\frac{\sin(2\phi(\mu^n+1))}{4(\mu^n+1)}\right.\\
&\qquad\left.+\frac{\sin(2\phi(\mu^n-1))}{4(\mu^n-1)}\right]_0^{\pi}\\
&= \frac{1}{4}.
\end{aligned}
\tag{6.73}
$$

Similarly it can be shown that

$$
\mathrm{E}[x_j^2]=\mathrm{E}[x_k^2]=\int_{-\infty}^{\infty}\rho(x)x^2dx=\int_{-1}^{1}\frac{1}{\pi\sqrt{1-x^2}}\,x^2dx=\frac{1}{2}. \tag{6.74}
$$

Putting (6.73) and (6.74) into (6.71), it is proved that the autovariance for $\{x_k^2\}$ is vanishing for the Chebyshev map of degree larger than one.

For the case where $j=k$,

$$
\begin{aligned}
\mathrm{cov}[x_k^2,x_k^2] &= \mathrm{var}[x_k^2]\\
&= \mathrm{E}[x_k^4]-\mathrm{E}^2[x_k^2]\\
&= \int_{-1}^{1}\frac{1}{\pi\sqrt{1-x^2}}\,x^4\,dx-\frac{1}{4}\\
&= \int_{\pi}^{0}\frac{1}{\pi\sin\phi}\cos^4\phi\,(-\sin\phi)\,d\phi-\frac{1}{4}\\
&= \frac{1}{8\pi}\int_0^{\pi}(3+4\cos 2\phi+\cos 4\phi)\,d\phi-\frac{1}{4}\\
&= \frac{1}{8}.
\end{aligned}
\tag{6.75}
$$

Derivation of $\mathbf{E}[x_j x_k^2]$

When $j=k$,

$$
\mathrm{E}[x_j x_k^2]=\mathrm{E}[x_k^3]=\int_{-\infty}^{\infty}x^3\rho(x)dx=\int_{-1}^{1}x^3\rho(x)dx=0 \tag{6.76}
$$

because $\rho(x)$ is an even function whereas x^3 is odd. Next, we consider the case where $j<k$. Assume $k=j+n$ for some positive integer n and we obtain

$$
\begin{aligned}
\mathrm{E}[x_j x_k^2] &= \int_{-\infty}^{\infty}\rho(x)\,x\,(h^{(n)}(x))^2dx\\
&= \int_{-1}^{1}\frac{1}{\pi\sqrt{1-x^2}}\,x\,(h^{(n)}(x))^2dx\\
&= \frac{1}{\pi}\int_0^{\pi}\cos\phi\;(h^{(n)}(\cos\phi))^2\,d\phi
\end{aligned}
$$

$$
\begin{aligned}
&= \frac{1}{\pi}\int_0^{\pi} \cos\phi \ \cos^2(\mu^n\phi)\ d\phi \\
&= 0
\end{aligned}
\tag{6.77}
$$

in which the substitution $x = \cos\phi$ has been made. Thus, it can be concluded that $\mathrm{E}[x_j x_k^2] = 0$ for $j < k$. Finally, for $j > k$, we assume $j = k + m$ for some positive integer m. Making the substitution $x = \cos\phi$ and applying (6.68), we have

$$
\begin{aligned}
\mathrm{E}[x_{k+m}x_k^2] &= \int_{-\infty}^{\infty} \rho(x)\ x^2\ h^{(m)}(x)dx \\
&= \frac{1}{\pi}\int_0^{\pi} \cos^2\phi\ \cos(\mu^m\phi)\ d\phi \\
&= \frac{1}{2\pi}\int_0^{\pi} \left(\cos(\mu^m\phi) + \frac{\cos(\mu^m+2)\phi}{2} + \frac{\cos(\mu^m-2)\phi}{2}\right) d\phi \\
&= \frac{1}{4\pi}\int_0^{\pi} \cos(\mu^m - 2)\phi\ d\phi \\
&= \begin{cases} 0 & \text{for } m \geq 2 \text{ and all } \mu \geq 2 \\ 0 & \text{for } m = 1 \text{ and } \mu > 2 \\ \frac{1}{4} & \text{for } m = 1 \text{ and } \mu = 2. \end{cases}
\end{aligned}
\tag{6.78}
$$

Thus, we conclude that for the Chebyshev map of degree greater than one,

$$
\mathrm{E}[x_j x_k^2] = 0 \quad \text{for } \ j - k \neq 1. \tag{6.79}
$$

7. Coexistence of Chaos-Based and Conventional Narrowband Digital Communication Systems

Chaos-based communication systems represent a new category of spread-spectrum communication systems whose working principle differs significantly from conventional direct-sequence and frequency-hopping spread-spectrum systems. However, like all other kinds of spread-spectrum systems, chaos-based systems are required to provide reasonable bit error performance in the presence of a narrowband signal which can be generated from an *intruder* or a coexisting conventional communication system. In particular, the frequency band of this foreign narrowband signal can fall within the bandwidth of the chaos-based system in question. Such a scenario may occur in normal practice when chaos-based systems are introduced while the conventional systems are still in operation. It is therefore important to examine the coexistence of chaos-based and conventional systems. The objective of this chapter is to evaluate the performance of the chaos-based system when its bandwidth overlaps with that of a coexisting conventional system. In particular, the chaos-based systems under study are the coherent chaos-shift-keying (CSK) system and the non-coherent differential CSK (DCSK) system, whereas the conventional system used in the study employs the standard binary phase-shift-keying (BPSK) scheme [Proakis and Salehi (1994), Ziemer and Tranter (1995)].

7.1 Overview of the Problem

The basic problem considered in this chapter is the coexistence of chaos-based systems and conventional systems. Specifically, we are interested in finding the performance of a chaos-based system and the extent to which it is affected by the presence of a conventional narrowband system whose bandwidth falls within that of the chaos-based system in question. This scenario has practical significance, as can be easily appreciated when one considers the introduction of chaos-based communication systems while conventional systems are still in operation. When it happens, chaos-based systems and conventional systems are actually interfering with one another. The ability of a chaos-based communication system to coexist with a conventional communication system is therefore an important issue that should be thoroughly investigated. The main questions are whether the interference can be tolerated and under

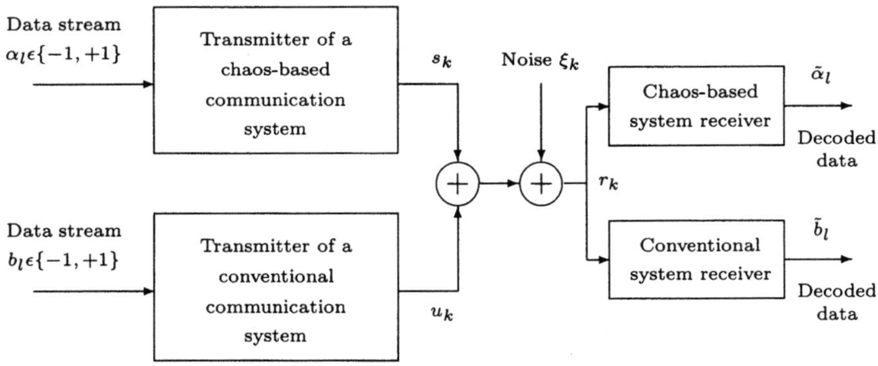

Fig. 7.1. Block diagram of a combined chaos-based-conventional digital communication system.

what conditions both kinds of systems will operate with satisfactory performances. To answer these questions, we first present an analytical method for evaluating the performances of the chaos-based communication system and the conventional system when their corresponding bandwidths overlap substantially. Then, based on the analytical bit error rates (BERs), we evaluate the coexistence performance for a range of noise levels, power ratios and spreading factors.

7.2 System Description

We consider a chaos-based communication system and a conventional system whose bandwidths overlap significantly. We refer to the whole system as *combined chaos-based-conventional system,* which can be represented by the block diagram shown in Fig. 7.1. In this system, two independent data streams are assumed to be sent at the same data rate. Our analysis will proceed in a discrete-time fashion. At time k, denote the output of the chaos transmitter by s_k and that of the conventional transmitter by u_k. These two signals are then added, as well as corrupted by noise ξ_k in the channel, before they arrive at the receiving end. At the receiver, based on the incoming signal r_k, the receivers of the chaos-based system and the conventional system will attempt to recover their respective data streams. Coherent or non-coherent detection schemes may be applied in the receivers, depending upon the modulation methods used in the transmitter.

Clearly, the signals from the chaotic and the conventional systems will be interfering with each other and thus the performance of each system will be degraded. Specifically, we will consider a "combined CSK-BPSK" system and a "combined DCSK-BPSK" system, and will attempt to develop analytical expressions for the bit error rates of the recovered data streams.

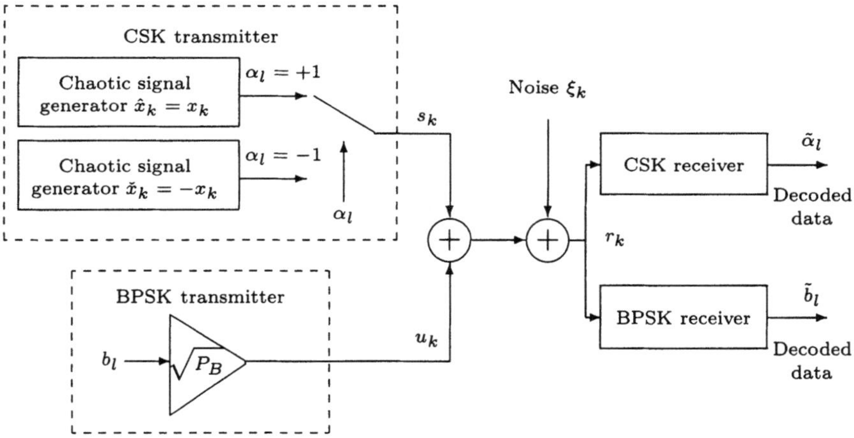

Fig. 7.2. Block diagram of a combined CSK-BPSK communication system.

7.3 Performance Analysis of Combined CSK-BPSK System

We first consider a discrete-time baseband equivalent model of a combined CSK-BPSK communication system, as shown in Fig. 7.2. We assume that the CSK system and the BPSK system have identical bit rate and that their bit streams are synchronized. Also, the carrier frequencies of the two systems are identical and synchronized. Further, "−1" and "+1" occur with equal probabilities in the bit streams of both systems.

Generally, in the CSK transmitter, a pair of chaotic sequences, denoted by $\{\hat{x}_k\}$ and $\{\check{x}_k\}$, are generated by two chaotic maps. If the symbol "+1" is sent, $\{\hat{x}_k\}$ is transmitted during a bit period, and if "−1" is sent, $\{\check{x}_k\}$ is transmitted. For simplicity, we consider here a CSK system in which one chaos generator is used to produce chaotic signal samples $\{x_k\}$ for $k = 1, 2, \ldots$. The two possible transmitted sequences are $\{\hat{x}_k = x_k\}$ and $\{\check{x}_k = -x_k\}$. Suppose $\alpha_l \in \{-1, +1\}$ is the symbol to be sent during the lth bit period. Define the spreading factor, 2β, as the number of chaotic samples used to transmit one binary symbol. During the lth bit duration, i.e., for $k = 2\beta(l-1)+1, 2\beta(l-1)+2, \ldots, 2\beta l$, the output of the CSK transmitter is

$$s_k = \alpha_l x_k. \tag{7.1}$$

In the BPSK system, we denote the lth transmitted symbol by $b_l \in \{-1, +1\}$. Moreover, the signal power of the narrowband equivalent signal is P_B. Thus, during the lth bit duration, i.e., for $k = 2\beta(l-1)+1, 2\beta(l-1)+2, \ldots, 2\beta l$, the transmitted signal is constant and is represented by

$$u_k = \sqrt{P_B} b_l. \tag{7.2}$$

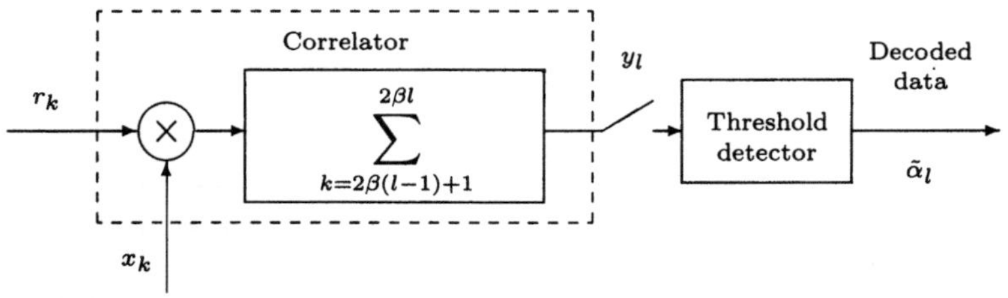

Fig. 7.3. Block diagram of a coherent CSK receiver.

The CSK and BPSK signals are combined and corrupted by an additive white Gaussian noise in the channel, before arriving at the receiving end. Thus, the received signal, denoted by r_k, is given by

$$r_k = s_k + u_k + \xi_k \tag{7.3}$$

where ξ_k is a Gaussian noise sample of zero mean and variance (power spectral density) $N_0/2$. For each of the CSK and BPSK receivers, we will consider the lth bit and derive the error probability over all transmitted bits.

7.3.1 Performance of the CSK System in Combined CSK-BPSK System

Assume that a correlator-type receiver is employed. Referring to Fig. 7.3, the correlator output for the lth bit, y_l, is given by

$$\begin{aligned} y_l &= \sum_{k=2\beta(l-1)+1}^{2\beta l} r_k x_k \\ &= \underbrace{\alpha_l \sum_{k=2\beta(l-1)+1}^{2\beta l} x_k^2}_{\text{required signal}} + \underbrace{\sqrt{P_B} b_l \sum_{k=2\beta(l-1)+1}^{2\beta l} x_k}_{\text{interfering BPSK signal}} \\ &\quad + \underbrace{\sum_{k=2\beta(l-1)+1}^{2\beta l} \xi_k x_k}_{\text{noise}} . \end{aligned} \tag{7.4}$$

Suppose a "+1" is transmitted in both CSK and BPSK systems during the lth symbol duration, i.e., $\alpha_l = +1$ and $b_l = +1$. For simplicity we write $y_l \mid (\alpha_l = +1, b_l = +1)$ as

$$y_l \mid (\alpha_l = +1, b_l = +1) = A + B + C \tag{7.5}$$

where A, B and C are the required signal, interfering BPSK signal and noise, respectively, and are defined as

$$A = \sum_{k=2\beta(l-1)+1}^{2\beta l} x_k^2 \tag{7.6}$$

$$B = \sqrt{P_B} \sum_{k=2\beta(l-1)+1}^{2\beta l} x_k \tag{7.7}$$

$$C = \sum_{k=2\beta(l-1)+1}^{2\beta l} \xi_k x_k. \tag{7.8}$$

The mean of $y_l \mid (\alpha_l = +1, b_l = +1)$ is

$$\begin{aligned}
\mathrm{E}[y_l \mid (\alpha_l = +1, b_l = +1)] &= \mathrm{E}[A] + \mathrm{E}[B] + \mathrm{E}[C] \\
&= \mathrm{E}\left[\sum_{k=2\beta(l-1)+1}^{2\beta l} x_k^2\right] \\
&\quad + \mathrm{E}\left[\sqrt{P_B} \sum_{k=2\beta(l-1)+1}^{2\beta l} x_k\right] \\
&\quad + \mathrm{E}\left[\sum_{k=2\beta(l-1)+1}^{2\beta l} \xi_k x_k\right] \\
&= \sum_{k=2\beta(l-1)+1}^{2\beta l} \mathrm{E}[x_k^2] + \sqrt{P_B} \sum_{k=2\beta(l-1)+1}^{2\beta l} \mathrm{E}[x_k] \\
&\quad + \sum_{k=2\beta(l-1)+1}^{2\beta l} \mathrm{E}[\xi_k]\mathrm{E}[x_k] \\
&= 2\beta P_s + 2\beta\sqrt{P_B}\mathrm{E}[x_k]
\end{aligned} \tag{7.9}$$

where $P_s = \mathrm{E}[x_k^2]$ denotes the *average power* of the chaotic signal. The last equality holds because $\mathrm{E}[\xi_k] = 0$. The variance of $y_l \mid (\alpha_l = +1, b_l = +1)$ is

$$\begin{aligned}
\mathrm{var}[y_l \mid (\alpha_l = +1, b_l = +1)] &= \mathrm{var}[A] + \mathrm{var}[B] + \mathrm{var}[C] \\
&\quad + 2\mathrm{cov}[A, B] + 2\mathrm{cov}[B, C] \\
&\quad + 2\mathrm{cov}[A, C]
\end{aligned} \tag{7.10}$$

where $\mathrm{cov}[X, Y]$ is the covariance of X and Y defined as

$$\mathrm{cov}[X, Y] = \mathrm{E}[XY] - \mathrm{E}[X]\mathrm{E}[Y]. \tag{7.11}$$

It can be proved that both $\mathrm{cov}[A, C]$ and $\mathrm{cov}[B, C]$ are zero (see Appendix 7A). Hence, (7.10) can be simplified to

$$\mathrm{var}[y_l \mid (\alpha_l = +1, b_l = +1)] = \mathrm{var}[A] + \mathrm{var}[B] + \mathrm{var}[C] + 2\mathrm{cov}[A, B]. \tag{7.12}$$

The mean value and the average power of the chaotic signal can be computed by numerical simulation. If the invariant probability density function of $\{x_k\}$

is available, in most cases, the mean and the average power can be obtained not only by numerical integration, but also in analytical forms. The variance and covariance terms in (7.12) can also be computed using the aforementioned techniques. Hence, $\mathrm{E}[y_l \mid (\alpha_l = +1, b_l = +1)]$ and $\mathrm{var}[y_l \mid (\alpha_l = +1, b_l = +1)]$ can be evaluated.

For the lth symbol, an error occurs when $y_l \leq 0 \mid (\alpha_l = +1, b_l = +1)$. Since $y_l \mid (\alpha_l = +1, b_l = +1)$ is the sum of a large number of random variables, we may assume that it follows a normal distribution. The error probability is thus given by

$$\begin{aligned} &\mathrm{Prob}(y_l \leq 0 \mid (\alpha_l = +1, b_l = +1)) \\ &\quad = \frac{1}{2}\mathrm{erfc}\left(\frac{\mathrm{E}\left[y_l \mid (\alpha_l = +1, b_l = +1)\right]}{\sqrt{2\mathrm{var}\left[y_l \mid (\alpha_l = +1, b_l = +1)\right]}}\right) \end{aligned} \tag{7.13}$$

where erfc(.) is the complementary error function defined as

$$\mathrm{erfc}(\psi) \equiv \frac{2}{\sqrt{\pi}} \int_{\psi}^{\infty} e^{-\lambda^2} d\lambda. \tag{7.14}$$

Similarly, when $\alpha_l = +1$ and $b_l = -1$, the output of the correlator can be shown equal to

$$y_l \mid (\alpha_l = +1, b_l = -1) = A - B + C. \tag{7.15}$$

Likewise, the mean and variance of $y_l \mid (\alpha_l = +1, b_l = -1)$, denoted by $\mathrm{E}[y_l \mid (\alpha_l = +1, b_l = -1)]$ and $\mathrm{var}[y_l \mid (\alpha_l = +1, b_l = -1)]$, can be found as

$$\mathrm{E}[y_l \mid (\alpha_l = +1, b_l = -1)] = 2\beta P_s - 2\beta\sqrt{P_B}\mathrm{E}[x_k] \tag{7.16}$$

$$\begin{aligned} \mathrm{var}[y_l \mid (\alpha_l = +1, b_l = -1)] &= \mathrm{var}[A] + \mathrm{var}[B] + \mathrm{var}[C] \\ &\quad - 2\mathrm{cov}[A, B] \end{aligned} \tag{7.17}$$

where A, B, and C are defined in (7.6) to (7.8). The corresponding error probability is

$$\begin{aligned} &\mathrm{Prob}(y_l \leq 0 \mid (\alpha_l = +1, b_l = -1)) \\ &\quad = \frac{1}{2}\mathrm{erfc}\left(\frac{\mathrm{E}\left[y_l \mid (\alpha_l = +1, b_l = -1)\right]}{\sqrt{2\mathrm{var}\left[y_l \mid (\alpha_l = +1, b_l = -1)\right]}}\right). \end{aligned} \tag{7.18}$$

Hence, for the CSK system, given a "+1" is sent during the lth bit duration, the error probability is given by

$$\begin{aligned} \mathrm{BER}^{(l)}_{\text{CSK-I}} &= \mathrm{Prob}(b_l = +1) \times \mathrm{Prob}(y_l \leq 0 \mid (\alpha_l = +1, b_l = +1) \\ &\quad + \mathrm{Prob}(b_l = -1) \times \mathrm{Prob}(y_l \leq 0 \mid (\alpha_l = +1, b_l = -1) \\ &= \frac{1}{4}\left[\mathrm{erfc}\left(\frac{\mathrm{E}[y_l \mid (\alpha_l = +1, b_l = +1)]}{\sqrt{2\mathrm{var}[y_l \mid (\alpha_l = +1, b_l = +1)]}}\right)\right. \end{aligned}$$

$$+ \operatorname{erfc}\left(\frac{\mathrm{E}[y_l \mid (\alpha_l = +1, b_l = -1)]}{\sqrt{2\mathrm{var}[y_l \mid (\alpha_l = +1, b_l = -1)]}}\right)\Bigg]. \quad (7.19)$$

Also, given "−1" is sent during the lth symbol duration in the CSK system, i.e., $\alpha_l = -1$, it can be shown that

$$\begin{aligned}
\mathrm{E}[y_l \mid (\alpha_l = -1, b_l = +1)] &= -2\beta P_s + 2\beta\sqrt{P_B}\mathrm{E}[x_k] & (7.20)\\
\mathrm{var}[y_l \mid (\alpha_l = -1, b_l = +1)] &= \mathrm{var}[A] + \mathrm{var}[B] + \mathrm{var}[C] \\
&\quad - 2\mathrm{cov}[A, B] & (7.21)\\
\mathrm{E}[y_l \mid (\alpha_l = -1, b_l = -1)] &= -2\beta P_s - 2\beta\sqrt{P_B}\mathrm{E}[x_k] & (7.22)\\
\mathrm{var}[y_l \mid (\alpha_l = -1, b_l = -1)] &= \mathrm{var}[A] + \mathrm{var}[B] + \mathrm{var}[C] \\
&\quad + 2\mathrm{cov}[A, B] & (7.23)
\end{aligned}$$

where A, B, and C are defined in (7.6) to (7.8). The error probability, given a "−1" is sent, is then equal to

$$\begin{aligned}
\mathrm{BER}^{(l)}_{\text{CSK-II}} &= \mathrm{Prob}(b_l = +1) \times \mathrm{Prob}(y_l > 0 \mid (\alpha_l = -1, b_l = +1) \\
&\quad + \mathrm{Prob}(b_l = -1) \times \mathrm{Prob}(y_l > 0 \mid (\alpha_l = -1, b_l = -1) \\
&= \frac{1}{4}\left[\operatorname{erfc}\left(\frac{-\mathrm{E}[y_l \mid (\alpha_l = -1, b_l = +1)]}{\sqrt{2\mathrm{var}[y_l \mid (\alpha_l = -1, b_l = +1)]}}\right)\right. \\
&\quad \left. + \operatorname{erfc}\left(\frac{-\mathrm{E}[y_l \mid (\alpha_l = -1, b_l = -1)]}{\sqrt{2\mathrm{var}[y_l \mid (\alpha_l = -1, b_l = -1)]}}\right)\right]. \qquad (7.24)
\end{aligned}$$

Hence, the overall error probability of the lth transmitted symbol is

$$\begin{aligned}
\mathrm{BER}^{(l)}_{\text{CSK}} &= \mathrm{Prob}(\alpha_l = +1) \times \mathrm{BER}^{(l)}_{\text{CSK-I}} + \mathrm{Prob}(\alpha_l = -1) \times \mathrm{BER}^{(l)}_{\text{CSK-II}} \\
&= \frac{1}{2}\left[\mathrm{BER}^{(l)}_{\text{CSK-I}} + \mathrm{BER}^{(l)}_{\text{CSK-II}}\right]. \qquad (7.25)
\end{aligned}$$

It can be seen from (7.19), (7.24) and (7.25) that $\mathrm{BER}^{(l)}_{\text{CSK}}$ is independent of l. Thus, the error probability of the lth transmitted symbol is the same as the BER of the system. In the combined CSK-BPSK system, the BER of the symbols carried by the CSK signal, denoted by $\mathrm{BER}_{\text{CSK}}$, is therefore

$$\mathrm{BER}_{\text{CSK}} = \mathrm{BER}^{(l)}_{\text{CSK}} = \frac{1}{2}[\mathrm{BER}^{(l)}_{\text{CSK-I}} + \mathrm{BER}^{(l)}_{\text{CSK-II}}]. \qquad (7.26)$$

Hence, (7.19) and (7.24) can be computed and substituted into (7.26) to obtain the BER of the system.

At this point, we make a few assumptions in order to further simplify the analysis. These assumptions can be easily justified for the chaotic sequences generated by the logistic map and by all Chebyshev maps of degree larger than one.

1. The mean value of $\{x_k\}$ is zero. The justification for this assumption is that no power should be wasted in sending non-information-bearing dc component through the channel. The condition also optimizes the performance of the joint CSK-BPSK scheme because it ensures that the chaotic sequences are restricted to the plane orthogonal to the basis vector $[1, 1, 1, 1, 1, 1, \ldots]$ used for BPSK. In practice, any dc component generated by the chaos generator can be removed artificially before transmission.
2. The autovariance of $\{x_k\}$ vanishes.
3. The autovariance of $\{x_k^2\}$ vanishes.
4. The correlation of $\{x_k\}$ and $\{x_m^2\}$ vanishes for $k - m \neq 1$.

The above assumptions can be translated to

$$\mathrm{E}[x_k] = 0 \tag{7.27}$$

$$\mathrm{cov}[x_k, x_m] = \mathrm{E}[x_k x_m] - \mathrm{E}[x_k]\mathrm{E}[x_m] = 0 \quad \text{for} \quad k \neq m \tag{7.28}$$

$$\mathrm{cov}[x_k^2, x_m^2] = \mathrm{E}[x_k^2 x_m^2] - \mathrm{E}[x_k^2]\mathrm{E}[x_m^2] = 0 \quad \text{for} \quad k \neq m \tag{7.29}$$

$$\mathrm{E}[x_k x_m^2] = 0 \quad \text{for} \quad k - m \neq 1. \tag{7.30}$$

Thus, (7.9), (7.16), (7.20) and (7.22) become

$$\begin{aligned}\mathrm{E}[y_l \mid (\alpha_l = +1, b_l = +1)] &= \mathrm{E}[y_l \mid (\alpha_l = +1, b_l = -1)] \\ &= 2\beta P_s \end{aligned} \tag{7.31}$$

$$\begin{aligned}\mathrm{E}[y_l \mid (\alpha_l = -1, b_l = +1)] &= \mathrm{E}[y_l \mid (\alpha_l = -1, b_l = -1)] \\ &= -2\beta P_s \end{aligned} \tag{7.32}$$

and the variances of the variables A, B and C, and the covariance between A and B are given by (see Appendix 7A)

$$\mathrm{var}[A] = 2\beta\Lambda \tag{7.33}$$

$$\mathrm{var}[B] = 2\beta P_B P_s \tag{7.34}$$

$$\mathrm{var}[C] = \beta N_0 P_s \tag{7.35}$$

$$\begin{aligned}\mathrm{cov}[A, B] &= \sqrt{P_B}(2\beta - 1)\mathrm{E}[x_{k+1}x_k^2] \\ &\approx 2\beta\sqrt{P_B}\Omega \quad \text{for large } 2\beta \end{aligned} \tag{7.36}$$

where

$$\Lambda = \mathrm{var}[x_k^2] \tag{7.37}$$

$$\Omega = \mathrm{E}[x_{k+1}x_k^2]. \tag{7.38}$$

Note that Λ denotes the variance of $\{x_k^2\}$ and is different from the average power of the chaotic signal, P_s. Hence, (7.12), (7.17), (7.21) and (7.23) can be put as

$$\begin{aligned}\mathrm{var}[y_l \mid (\alpha_l = +1, b_l = +1)] &= \mathrm{var}[y_l \mid (\alpha_l = -1, b_l = -1)] \\ &= 2\beta\Lambda + 2\beta P_B P_s + \beta N_0 P_s\end{aligned}$$

$$
\begin{aligned}
& \quad + 2(2\beta - 1)\sqrt{P_B}\Omega \\
& \approx \beta(2\Lambda + 2P_B P_s + N_0 P_s \\
& \quad + 4\sqrt{P_B}\Omega) \qquad (7.39) \\
\mathrm{var}[y_l \mid (\alpha_l = +1, b_l = -1)] &= \mathrm{var}[y_l \mid (\alpha_l = -1, b_l = +1)] \\
&= 2\beta\Lambda + 2\beta P_B P_s + \beta N_0 P_s \\
& \quad - 2(2\beta - 1)\sqrt{P_B}\Omega \\
& \approx \beta(2\Lambda + 2P_B P_s + N_0 P_s \\
& \quad - 4\sqrt{P_B}\Omega). \qquad (7.40)
\end{aligned}
$$

Substituting (7.31), (7.32), (7.39) and (7.40) into (7.19), (7.24) and (7.26), the BER can be found as

$$
\begin{aligned}
\mathrm{BER}_{\mathrm{CSK}} &\approx \frac{1}{4}\mathrm{erfc}\left(\frac{2\beta P_s}{\sqrt{2\beta(2\Lambda + 2P_B P_s + N_0 P_s + 4\sqrt{P_B}\Omega)}}\right) \\
& \quad + \frac{1}{4}\mathrm{erfc}\left(\frac{2\beta P_s}{\sqrt{2\beta(2\Lambda + 2P_B P_s + N_0 P_s - 4\sqrt{P_B}\Omega)}}\right) \\
&= \frac{1}{4}\mathrm{erfc}\left[\left(\frac{\Psi}{\beta} + \frac{P_B}{\beta P_s} + \frac{N_0}{2\beta P_s} + \frac{2\Phi}{\beta}\sqrt{\frac{P_B}{P_s}}\right)^{-\frac{1}{2}}\right] \\
& \quad + \frac{1}{4}\mathrm{erfc}\left[\left(\frac{\Psi}{\beta} + \frac{P_B}{\beta P_s} + \frac{N_0}{2\beta P_s} - \frac{2\Phi}{\beta}\sqrt{\frac{P_B}{P_s}}\right)^{-\frac{1}{2}}\right] \qquad (7.41) \\
&= \frac{1}{4}\mathrm{erfc}\left[\left(\frac{\Psi}{\beta} + \frac{P_B}{\beta P_s} + \left(\frac{E_b}{N_0}\right)^{-1} + \frac{2\Phi}{\beta}\sqrt{\frac{P_B}{P_s}}\right)^{-\frac{1}{2}}\right] \\
& \quad + \frac{1}{4}\mathrm{erfc}\left[\left(\frac{\Psi}{\beta} + \frac{P_B}{\beta P_s} + \left(\frac{E_b}{N_0}\right)^{-1} - \frac{2\Phi}{\beta}\sqrt{\frac{P_B}{P_s}}\right)^{-\frac{1}{2}}\right] \\
& \qquad (7.42)
\end{aligned}
$$

where $E_b = 2\beta P_s$ denotes the average bit energy of the CSK system, and Ψ and Φ are given by

$$
\Psi = \frac{\mathrm{var}[x_k^2]}{P_s^2} \qquad (7.43)
$$

$$
\Phi = \frac{\mathrm{E}[x_{k+1}x_k^2]}{P_s^{3/2}} \qquad (7.44)
$$

which are dependent only upon the type of the chaotic sequence used (for Ψ, see Sect. 3.2.3.2 for details). The expression given above is thus the analytical BER for the noisy coherent CSK system in a combined communication environment. Note that for fixed BPSK signal power P_B and noise power spectral density $N_0/2$, the BER can be improved by making one or more of the following adjustments.

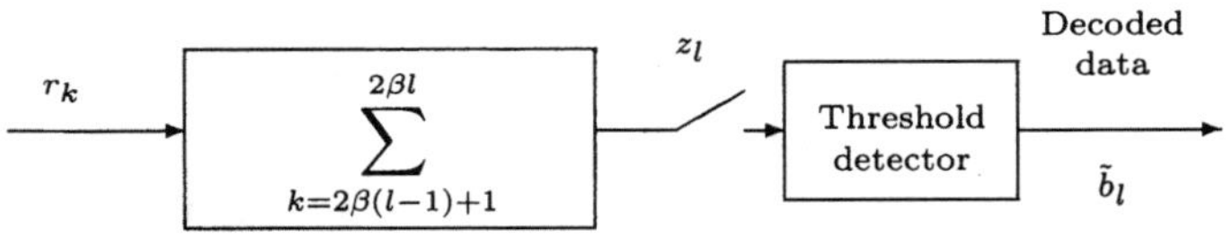

Fig. 7.4. Block diagram of a BPSK receiver.

1. Reduce the variance of $\{x_k^2\}$ for a given P_s.
2. Reduce the absolute value of Ω, i.e., $\mathrm{E}[x_{k+1}x_k^2]$, for a given P_s.
3. Increase the spreading factor 2β.
4. Increase the CSK signal power P_s.

In particular, when the BPSK signal power is zero, i.e., $P_B = 0$, it can be readily shown that the bit error rate reduces to (see Sect. 5.2.1)

$$\mathrm{BER}_{\mathrm{CSK}} \mid (P_B = 0) = \frac{1}{2}\mathrm{erfc}\left[\left(\frac{\Psi}{\beta} + \frac{N_0}{2\beta P_s}\right)^{-\frac{1}{2}}\right] \tag{7.45}$$

$$= \frac{1}{2}\mathrm{erfc}\left[\left(\frac{\Psi}{\beta} + \left(\frac{E_b}{N_0}\right)^{-1}\right)^{-\frac{1}{2}}\right]. \tag{7.46}$$

7.3.2 Performance of the BPSK System in Combined CSK-BPSK System

In the BPSK receiver shown in Fig. 7.4, the incoming signal samples within a symbol period are summed to give z_l, i.e.,

$$z_l = \sum_{k=2\beta(l-1)+1}^{2\beta l} r_k$$

$$= \underbrace{\alpha_l \sum_{k=2\beta(l-1)+1}^{2\beta l} x_k}_{\text{interfering CSK signal}} + \underbrace{2\beta\sqrt{P_B}b_l}_{\text{required signal}} + \underbrace{\sum_{k=2\beta(l-1)+1}^{2\beta l} \xi_k}_{\text{noise}}. \tag{7.47}$$

Using similar procedures as in the previous section, it can be shown that for a given b_l, the mean and variance of z_l, denoted by $\mathrm{E}[z_l \mid b_l]$ and $\mathrm{var}[z_l \mid b_l]$, respectively, are given by

$$\mathrm{E}[z_l \mid b_l] = 2\beta\alpha_l \mathrm{E}[x_k] + 2\beta\sqrt{P_B}b_l \tag{7.48}$$

$$\mathrm{var}[z_l \mid b_l] = \mathrm{var}\left[\sum_{k=2\beta(l-1)+1}^{2\beta l} x_k\right] + \beta N_0. \tag{7.49}$$

Assuming that (i) the mean value of $\{x_k\}$ is zero, and (ii) the autovariance of $\{x_k\}$ vanishes, putting (7.27) and (7.28) into (7.48) and (7.49) gives

$$\begin{aligned} \mathrm{E}[z_l \mid b_l] &= 2\beta\sqrt{P_B}b_l && (7.50)\\ \mathrm{var}[z_l \mid b_l] &= 2\beta P_s + \beta N_0. && (7.51) \end{aligned}$$

Suppose $b_l = +1$. Since $z_l \mid (b_l = +1)$ is the sum of a large number of random variables, we may assume that it is normally distributed. An error occurs when $z_l \leq 0 \mid (b_l = +1)$, and the corresponding error probability is given by

$$\mathrm{Prob}(z_l \leq 0 \mid (b_l = +1)) = \frac{1}{2}\mathrm{erfc}\left(\frac{\mathrm{E}\left[z_l \mid (b_l = +1)\right]}{\sqrt{2\mathrm{var}\left[z_l \mid (b_l = +1)\right]}}\right). \tag{7.52}$$

Likewise, given $b_l = -1$, the error probability is

$$\mathrm{Prob}(z_l > 0 \mid (b_l = -1)) = \frac{1}{2}\mathrm{erfc}\left(\frac{-\mathrm{E}\left[z_l \mid (b_l = -1)\right]}{\sqrt{2\mathrm{var}\left[z_l \mid (b_l = -1)\right]}}\right). \tag{7.53}$$

Putting (7.50) and (7.51) into (7.52) and (7.53), the error probability for the lth transmitted BPSK symbol can be found as

$$\begin{aligned} \mathrm{BER}^{(l)}_{\text{BPSK-CSK}} &= \mathrm{Prob}(b_l = +1) \times \mathrm{Prob}(z_l \leq 0 \mid (b_l = +1))\\ &\quad + \mathrm{Prob}(b_l = -1) \times \mathrm{Prob}(z_l > 0 \mid (b_l = -1))\\ &= \frac{1}{4}\mathrm{erfc}\left(\frac{\mathrm{E}[z_l \mid (b_l = +1)]}{\sqrt{2\mathrm{var}[z_l \mid (b_l = +1)]}}\right)\\ &\quad + \frac{1}{4}\mathrm{erfc}\left(\frac{-\mathrm{E}[z_l \mid (b_l = -1)]}{\sqrt{2\mathrm{var}[z_l \mid (b_l = -1)]}}\right)\\ &= \frac{1}{2}\mathrm{erfc}\left(\frac{2\beta\sqrt{P_B}}{\sqrt{4\beta P_s + 2\beta N_0}}\right). && (7.54) \end{aligned}$$

Since $\mathrm{BER}^{(l)}_{\text{BPSK-CSK}}$ is independent of l, the error probability of the lth transmitted symbol is the same as the BER of the system. Therefore, the BER of the BPSK system, denoted by $\mathrm{BER}_{\text{BPSK-CSK}}$, is

$$\begin{aligned} \mathrm{BER}_{\text{BPSK-CSK}} &= \mathrm{BER}^{(l)}_{\text{BPSK-CSK}}\\ &= \frac{1}{2}\mathrm{erfc}\left(\frac{2\beta\sqrt{P_B}}{\sqrt{4\beta P_s + 2\beta N_0}}\right)\\ &= \frac{1}{2}\mathrm{erfc}\left[\left(\frac{P_s}{\beta P_B} + \frac{N_0}{2\beta P_B}\right)^{-\frac{1}{2}}\right] && (7.55)\\ &= \frac{1}{2}\mathrm{erfc}\left(\sqrt{\frac{E_B}{\tilde{N}_0}}\right) && (7.56) \end{aligned}$$

where

$$E_B = 2\beta P_B \tag{7.57}$$

denotes the bit energy of the BPSK signal and

$$\frac{\tilde{N_0}}{2} = \frac{N_0}{2} + P_s \tag{7.58}$$

represents the equivalent noise power spectral density when the interfering CSK signal is taken into consideration. Thus, the interfering CSK signal simply raises the noise level of the BPSK signal. The expression given in (7.55) or (7.56) is the analytical BER for the noisy coherent BPSK system in a combined communication environment. For a fixed chaotic signal power P_s, the BER can be improved by increasing the spreading factor 2β and/or increasing the BPSK signal power P_B.

Example — Consider the case where a logistic map is used for chaos generation. The form of the map is

$$x_{k+1} = g(x_k) = 1 - 2x_k^2. \tag{7.59}$$

The invariant probability density function of $\{x_k\}$, denoted by $\rho(x)$, is

$$\rho(x) = \begin{cases} \dfrac{1}{\pi\sqrt{1-x^2}} & \text{if } |x| < 1 \\ 0 & \text{otherwise.} \end{cases} \tag{7.60}$$

Using (7.60), it can be shown (see Appendix 3B and Appendix 7B) that the assumptions corresponding to (7.27) through (7.30) made earlier in Sect. 7.3.1 and the two assumptions made in Sect. 7.3.2 are all well justified for the chaotic sequences generated by the logistic map.[1] Moreover, we have

$$P_s = \mathrm{E}[x_k^2] = \int_{-\infty}^{\infty} x^2\rho(x)dx = \int_{-1}^{1} x^2\rho(x)dx = \frac{1}{2} \tag{7.61}$$

$$\Lambda = \mathrm{var}[x_k^2] = \mathrm{E}[x_k^4] - \mathrm{E}^2[x_k^2] = \int_{-1}^{1} x^4\rho(x)dx - \frac{1}{4} = \frac{1}{8} \tag{7.62}$$

$$\Omega = \mathrm{E}[x_{k+1}x_k^2] = -\frac{1}{4}. \tag{7.63}$$

For the case where the logistic map is used to generate the chaotic samples, we substitute (7.61) to (7.63) into (7.41) to obtain the BER of the CSK system, i.e.,

$$\mathrm{BER}_{\mathrm{CSK}} \approx \frac{1}{4}\mathrm{erfc}\left(\sqrt{\frac{2\beta}{1+4P_B+2N_0-4\sqrt{P_B}}}\right)$$

[1] In Appendix 6C, it has been shown that (7.27) to (7.30) are satisfied by the chaotic sequences generated by the class of Chebyshev maps of degree larger than one.

$$+\frac{1}{4}\text{erfc}\left(\sqrt{\frac{2\beta}{1+4P_B+2N_0+4\sqrt{P_B}}}\right). \tag{7.64}$$

Moreover, for the BPSK system, we put (7.61) into (7.55) to obtain

$$\text{BER}_{\text{BPSK-CSK}} = \frac{1}{2}\text{erfc}\left(\sqrt{\frac{2\beta P_B}{1+N_0}}\right). \tag{7.65}$$

7.4 Performance Analysis of Combined DCSK-BPSK System

In this section, we move on to a combined DCSK-BPSK system. In a DCSK system, the basic modulation process involves dividing the bit period into two equal slots. The first slot carries a reference chaotic signal, and the second slot bears the information. For a binary system, the second slot is the same copy or an inverted copy of the first slot depending upon the symbol sent being "+1" or "−1". This structural arrangement allows the detection to be done in a non-coherent manner requiring no reproduction of the same chaotic carrying signals at the receiver. Essentially, the detection of a DCSK signal can be accomplished by correlating the first and the second slots of the same symbol and comparing the correlator output with a threshold. Figure 7.5 shows the block diagram of a DCSK transmitter and receiver pair.

Making the same assumptions as in Sect. 7.3, we obtain the transmitted DCSK signal in the lth bit duration as

$$s_k = \begin{cases} x_k & \text{for } k = 2\beta(l-1)+1, 2\beta(l-1)+2, \ldots, 2\beta(l-1)+\beta \\ \alpha_l x_{k-\beta} & \text{for } k = 2\beta(l-1)+\beta+1, 2\beta(l-1)+\beta+2, \ldots, 2\beta l \end{cases} \tag{7.66}$$

and the BPSK signal as

$$u_k = \sqrt{P_B} b_l \text{ for } k = 2\beta(l-1)+1, 2\beta(l-1)+2, \ldots, 2\beta l. \tag{7.67}$$

All symbols and notations are as defined in the previous section. The noisy received signal, r_k, is given by

$$r_k = s_k + u_k + \xi_k. \tag{7.68}$$

7.4.1 Performance of the DCSK System in Combined DCSK-BPSK System

At the DCSK receiver, the detector essentially computes the correlation of the corrupted reference and data slots of the same symbol. We consider the output of the correlator for the lth received bit, y_l, which is given by

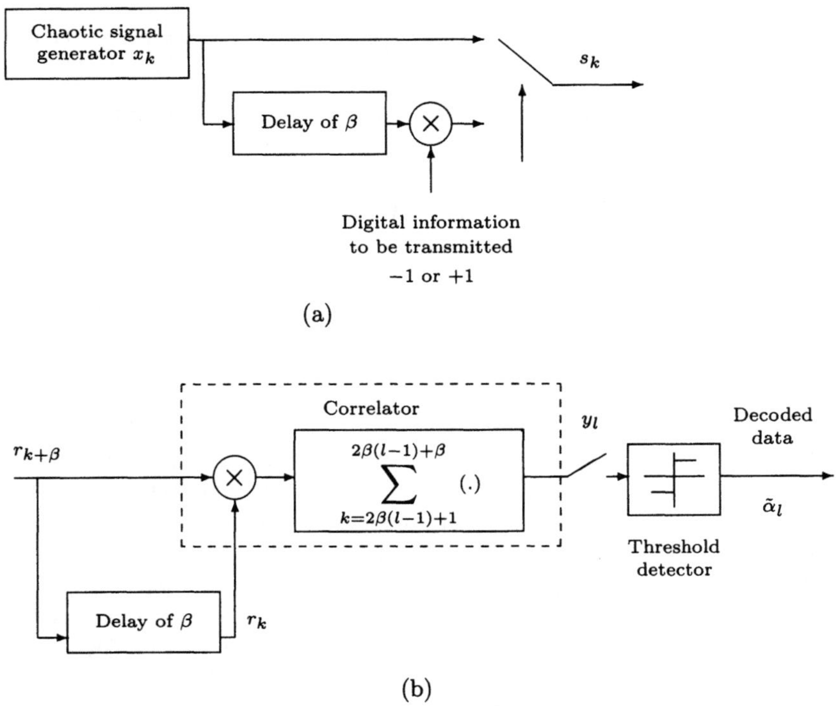

Fig. 7.5. Block diagram of a non-coherent DCSK system. (a) Transmitter; (b) receiver.

$$
\begin{aligned}
y_l &= \sum_{k=2\beta(l-1)+1}^{2\beta(l-1)+\beta} r_k r_{k+\beta} \\
&= \sum_{k=2\beta(l-1)+1}^{2\beta(l-1)+\beta} [x_k + u_k + \xi_k]\,[\alpha_l x_k + u_{k+\beta} + \xi_{k+\beta}] \\
&= \alpha_l \sum_{k=2\beta(l-1)+1}^{2\beta(l-1)+\beta} x_k^2 + \sqrt{P_B}b_l(1+\alpha_l) \sum_{k=2\beta(l-1)+1}^{2\beta(l-1)+\beta} x_k \\
&\quad + \beta P_B + \sqrt{P_B}b_l \sum_{k=2\beta(l-1)+1}^{2\beta(l-1)+\beta} [\xi_{k+\beta} + \xi_k] \\
&\quad + \sum_{k=2\beta(l-1)+1}^{2\beta(l-1)+\beta} x_k(\xi_{k+\beta} + \alpha_l \xi_k) + \sum_{k=2\beta(l-1)+1}^{2\beta(l-1)+\beta} \xi_k \xi_{k+\beta} \\
&= \underbrace{\alpha_l D}_{\text{required signal}} + \underbrace{\sqrt{P_B}b_l(1+\alpha_l)F + \beta P_B + \sqrt{P_B}b_l G}_{\text{interfering BPSK signal}}
\end{aligned}
$$

$$+ \underbrace{H + \alpha_l J + K}_{\text{noise}} \tag{7.69}$$

where

$$D = \sum_{k=2\beta(l-1)+1}^{2\beta(l-1)+\beta} x_k^2 \tag{7.70}$$

$$F = \sum_{k=2\beta(l-1)+1}^{2\beta(l-1)+\beta} x_k \tag{7.71}$$

$$G = \sum_{k=2\beta(l-1)+1}^{2\beta(l-1)+\beta} [\xi_{k+\beta} + \xi_k] \tag{7.72}$$

$$H = \sum_{k=2\beta(l-1)+1}^{2\beta(l-1)+\beta} x_k \xi_{k+\beta} \tag{7.73}$$

$$J = \sum_{k=2\beta(l-1)+1}^{2\beta(l-1)+\beta} x_k \xi_k \tag{7.74}$$

$$K = \sum_{k=2\beta(l-1)+1}^{2\beta(l-1)+\beta} \xi_k \xi_{k+\beta}. \tag{7.75}$$

Suppose a "+1" is transmitted in both DCSK and BPSK systems during the lth symbol duration, i.e., $\alpha_l = +1$ and $b_l = +1$. Then, (7.69) can be re-written as

$$y_l \mid (\alpha_l = +1, b_l = +1) = D + 2\sqrt{P_B}F + \beta P_B + \sqrt{P_B}G + H + J + K. \tag{7.76}$$

Similar to the case of the combined CSK-BPSK environment (Sect. 7.3.1), the mean and variance of $y_l \mid (\alpha_l = +1, b_l = +1)$ for the DCSK system can be evaluated by numerical simulations. Denote the respective mean and variance by $\mathrm{E}[y_l \mid (\alpha_l = +1, b_l = +1)]$ and $\mathrm{var}[y_l \mid (\alpha_l = +1, b_l = +1)]$. Since $y_l \mid (\alpha_l = +1, b_l = +1)$ is the sum of a large number of random variables, we may assume that it is normally distributed. An error occurs when $y_l \leq 0 \mid (\alpha_l = +1, b_l = +1)$, and the corresponding error probability is given by

$$\begin{aligned} &\mathrm{Prob}(y_l \leq 0 \mid (\alpha_l = +1, b_l = +1)) \\ &\quad = \frac{1}{2}\mathrm{erfc}\left(\frac{\mathrm{E}\left[y_l \mid (\alpha_l = +1, b_l = +1)\right]}{\sqrt{2\mathrm{var}\left[y_l \mid (\alpha_l = +1, b_l = +1)\right]}}\right). \end{aligned} \tag{7.77}$$

Likewise, for the case $\alpha_l = +1$ and $b_l = -1$, (7.69) becomes

$$y_l \mid (\alpha_l = +1, b_l = -1)$$

$$= D - 2\sqrt{P_B}F + \beta P_B - \sqrt{P_B}G + H + J + K. \qquad (7.78)$$

The corresponding error probability is

$$\begin{aligned} &\text{Prob}(y_l \leq 0 \mid (\alpha_l = +1, b_l = -1)) \\ &\qquad = \frac{1}{2}\text{erfc}\left(\frac{\text{E}\left[y_l \mid (\alpha_l = +1, b_l = -1)\right]}{\sqrt{2\text{var}\left[y_l \mid (\alpha_l = +1, b_l = -1)\right]}}\right). \end{aligned} \qquad (7.79)$$

Given a "+1" is sent by the DCSK signal in the lth symbol duration, the probability that an error occurs is equal to

$$\begin{aligned} \text{BER}^{(l)}_{\text{DCSK-I}} &= \text{Prob}(b_l = +1) \times \text{Prob}(y_l \leq 0 \mid (\alpha_l = +1, b_l = +1) \\ &\quad + \text{Prob}(b_l = -1) \times \text{Prob}(y_l \leq 0 \mid (\alpha_l = +1, b_l = -1) \\ &= \frac{1}{4}\left[\text{erfc}\left(\frac{\text{E}[y_l \mid (\alpha_l = +1, b_l = +1)]}{\sqrt{2\text{var}[y_l \mid (\alpha_l = +1, b_l = +1)]}}\right)\right. \\ &\quad \left. + \text{ erfc}\left(\frac{\text{E}[y_l \mid (\alpha_l = +1, b_l = -1)]}{\sqrt{2\text{var}[y_l \mid (\alpha_l = +1, b_l = -1)]}}\right)\right]. \end{aligned} \qquad (7.80)$$

Similarly, given that "−1" is sent during the lth symbol duration in the DCSK system, i.e., $\alpha_l = -1$, it can be shown that

$$\begin{aligned} y_l \mid (\alpha_l = -1, b_l = +1) &= -D + \beta P_B \\ &\quad + \sqrt{P_B}G + H - J + K \qquad (7.81) \\ y_l \mid (\alpha_l = -1, b_l = -1) &= -D + \beta P_B \\ &\quad - \sqrt{P_B}G + H - J + K \qquad (7.82) \end{aligned}$$

where D, G, H, J and K are defined in (7.70) to (7.75). Denote the respective means and variances of $y_l \mid (\alpha_l = -1, b_l = +1)$ and $y_l \mid (\alpha_l = -1, b_l = -1)$ by $\text{E}[y_l \mid (\alpha_l = -1, b_l = +1)]$, $\text{E}[y_l \mid (\alpha_l = -1, b_l = -1)]$, $\text{var}[y_l \mid (\alpha_l = -1, b_l = +1)]$ and $\text{var}[y_l \mid (\alpha_l = -1, b_l = -1)]$. The error probability, given a "−1" is sent, is then equal to

$$\begin{aligned} \text{BER}^{(l)}_{\text{DCSK-II}} &= \text{Prob}(b_l = +1) \times \text{Prob}(y_l > 0 \mid (\alpha_l = -1, b_l = +1) \\ &\quad + \text{Prob}(b_l = -1) \times \text{Prob}(y_l > 0 \mid (\alpha_l = -1, b_l = -1) \\ &= \frac{1}{4}\left[\text{erfc}\left(\frac{-\text{E}[y_l \mid (\alpha_l = -1, b_l = +1)]}{\sqrt{2\text{var}[y_l \mid (\alpha_l = -1, b_l = +1)]}}\right)\right. \\ &\quad \left. + \text{ erfc}\left(\frac{-\text{E}[y_l \mid (\alpha_l = -1, b_l = -1)]}{\sqrt{2\text{var}[y_l \mid (\alpha_l = -1, b_l = -1)]}}\right)\right]. \end{aligned} \qquad (7.83)$$

Since both (7.80) and (7.83) are independent of l, the bit error rate of the DCSK system under a combined communication environment, denoted by BER_{DCSK}, equals the overall error probability of the lth transmitted symbol ($\text{BER}^{(l)}_{\text{DCSK}}$), i.e.,

$$\begin{aligned}
\mathrm{BER}_{\mathrm{DCSK}} &= \mathrm{BER}^{(l)}_{\mathrm{DCSK}} \\
&= \mathrm{Prob}(\alpha_l = +1) \times \mathrm{BER}^{(l)}_{\mathrm{DCSK\text{-}I}} + \mathrm{Prob}(\alpha_l = -1) \times \mathrm{BER}^{(l)}_{\mathrm{DCSK\text{-}II}} \\
&= \frac{1}{2}[\mathrm{BER}^{(l)}_{\mathrm{DCSK\text{-}I}} + \mathrm{BER}^{(l)}_{\mathrm{DCSK\text{-}II}}]. \qquad (7.84)
\end{aligned}$$

To simplify the analysis, we make similar assumptions as in Sect. 7.3.1. With these assumptions, we apply (7.27)–(7.30) to (7.70)–(7.75) and obtain the relevant means, variances and covariances, i.e.,

$$\begin{array}{ll}
\mathrm{E}[D] = \beta\mathrm{E}[x_k^2] \equiv \beta P_s & \mathrm{var}[D] = \beta\mathrm{var}[x_k^2] \equiv \beta\Lambda \\
\mathrm{E}[F] = 0 & \mathrm{var}[F] = \beta P_s \\
\mathrm{E}[G] = 0 & \mathrm{var}[G] = \beta N_0 \\
\mathrm{E}[H] = 0 & \mathrm{var}[H] = \beta P_s N_0/2 \\
\mathrm{E}[J] = 0 & \mathrm{var}[J] = \beta P_s N_0/2 \\
\mathrm{E}[K] = 0 & \mathrm{var}[K] = \beta N_0^2/4 \\
\mathrm{cov}[D,F] = (\beta-1)\mathrm{E}[x_{k+1}x_k^2] \approx \beta\Omega & \mathrm{cov}[\chi,\gamma] = 0
\end{array} \qquad (7.85)$$

where

$$\chi, \gamma \in \{D, F, G, H, J, K\},$$

$$(\chi, \gamma) \neq (D, F) \ \text{or} \ (F, D),$$

and Λ and Ω are as defined in (7.37) and (7.38). Furthermore, it can be readily shown that

$$\begin{aligned}
\mathrm{E}[y_l \mid (\alpha_l = +1, b_l = +1)] &= \mathrm{E}[D] + 2\sqrt{P_B}\mathrm{E}[F] + \beta P_B \\
&\quad + \sqrt{P_B}\mathrm{E}[G] + \mathrm{E}[H] \\
&\quad + \mathrm{E}[J] + \mathrm{E}[K] \qquad (7.86) \\
\mathrm{E}[y_l \mid (\alpha_l = +1, b_l = -1)] &= \mathrm{E}[D] - 2\sqrt{P_B}\mathrm{E}[F] + \beta P_B \\
&\quad - \sqrt{P_B}\mathrm{E}[G] + \mathrm{E}[H] \\
&\quad + \mathrm{E}[J] + \mathrm{E}[K] \qquad (7.87) \\
\mathrm{E}[y_l \mid (\alpha_l = -1, b_l = +1)] &= -\mathrm{E}[D] + \beta P_B + \sqrt{P_B}\mathrm{E}[G] \\
&\quad + \mathrm{E}[H] - \mathrm{E}[J] + \mathrm{E}[K] \qquad (7.88) \\
\mathrm{E}[y_l \mid (\alpha_l = -1, b_l = -1)] &= -\mathrm{E}[D] + \beta P_B - \sqrt{P_B}\mathrm{E}[G] \\
&\quad + \mathrm{E}[H] - \mathrm{E}[J] + \mathrm{E}[K] \qquad (7.89) \\
\mathrm{var}[y_l \mid (\alpha_l = +1, b_l = +1)] &= \mathrm{var}[D] + 4P_B\mathrm{var}[F] + P_B\mathrm{var}[G] \\
&\quad + \mathrm{var}[H] + \mathrm{var}[J] + \mathrm{var}[K] \\
&\quad + 4\sqrt{P_B}\mathrm{cov}[D,F] \qquad (7.90) \\
\mathrm{var}[y_l \mid (\alpha_l = +1, b_l = -1)] &= \mathrm{var}[D] + 4P_B\mathrm{var}[F] + P_B\mathrm{var}[G] \\
&\quad + \mathrm{var}[H] + \mathrm{var}[J] + \mathrm{var}[K] \\
&\quad - 4\sqrt{P_B}\mathrm{cov}[D,F] \qquad (7.91) \\
\mathrm{var}[y_l \mid (\alpha_l = -1, b_l = +1)] &= \mathrm{var}[y_l \mid (\alpha_l = -1, b_l = -1)] \\
&= \mathrm{var}[D] + P_B\mathrm{var}[G] + \mathrm{var}[H] \\
&\quad + \mathrm{var}[J] + \mathrm{var}[K]. \qquad (7.92)
\end{aligned}$$

Putting (7.85) into (7.86) through (7.92), we obtain

$$\begin{aligned}
\mathrm{E}[y_l \mid (\alpha_l = +1, b_l = +1)] &= \mathrm{E}[y_l \mid (\alpha_l = +1, b_l = -1)] \\
&= \beta P_s + \beta P_B && (7.93) \\
\mathrm{E}[y_l \mid (\alpha_l = -1, b_l = +1)] &= \mathrm{E}[y_l \mid (\alpha_l = -1, b_l = -1)] \\
&= -\beta P_s + \beta P_B && (7.94) \\
\mathrm{var}[y_l \mid (\alpha_l = +1, b_l = +1)] &\approx \beta\Lambda + 4\beta P_B P_s + \beta P_B N_0 + \beta P_s N_0 \\
&\quad + \frac{\beta N_0^2}{4} + 4\beta\Omega\sqrt{P_B} && (7.95) \\
\mathrm{var}[y_l \mid (\alpha_l = +1, b_l = -1)] &\approx \beta\Lambda + 4\beta P_B P_s + \beta P_B N_0 + \beta P_s N_0 \\
&\quad + \frac{\beta N_0^2}{4} - 4\beta\Omega\sqrt{P_B} && (7.96) \\
\mathrm{var}[y_l \mid (\alpha_l = -1, b_l = +1)] &= \mathrm{var}[y_l \mid (\alpha_l = -1, b_l = -1)] \\
&= \beta\Lambda + \beta P_B N_0 + \beta P_s N_0 \\
&\quad + \frac{\beta N_0^2}{4}. && (7.97)
\end{aligned}$$

Also, putting (7.93)–(7.97) into (7.80), (7.83) and (7.84), we get the BER of the DCSK system, i.e.,

$$\begin{aligned}
&\mathrm{BER}_{\mathrm{DCSK}} \\
&\approx \frac{1}{8}\mathrm{erfc}\left(\frac{\beta P_s + \beta P_B}{\sqrt{2\beta\Lambda + 8\beta P_B P_s + 2\beta P_B N_0 + 2\beta P_s N_0 + \frac{\beta N_0^2}{2} + 8\beta\Omega\sqrt{P_B}}}\right) \\
&+ \frac{1}{8}\mathrm{erfc}\left(\frac{\beta P_s + \beta P_B}{\sqrt{2\beta\Lambda + 8\beta P_B P_s + 2\beta P_B N_0 + 2\beta P_s N_0 + \frac{\beta N_0^2}{2} - 8\beta\Omega\sqrt{P_B}}}\right) \\
&+ \frac{1}{4}\mathrm{erfc}\left(\frac{\beta P_s - \beta P_B}{\sqrt{2\beta\Lambda + 2\beta P_B N_0 + 2\beta P_s N_0 + \frac{\beta N_0^2}{2}}}\right) \qquad (7.98) \\
&= \frac{1}{8}\mathrm{erfc}\left(\left(1 + \frac{P_B}{P_s}\right)\left[\frac{2\Psi}{\beta} + \frac{8P_B}{\beta P_s} + \frac{4P_B}{P_s}\left(\frac{E_b}{N_0}\right)^{-1} + 4\left(\frac{E_b}{N_0}\right)^{-1}\right.\right. \\
&\qquad \left.\left. + 2\beta\left(\frac{E_b}{N_0}\right)^{-2} + \frac{8\Phi}{\beta}\sqrt{\frac{P_B}{P_s}}\right]^{-\frac{1}{2}}\right) \\
&+ \frac{1}{8}\mathrm{erfc}\left(\left(1 + \frac{P_B}{P_s}\right)\left[\frac{2\Psi}{\beta} + \frac{8P_B}{\beta P_s} + \frac{4P_B}{P_s}\left(\frac{E_b}{N_0}\right)^{-1} + 4\left(\frac{E_b}{N_0}\right)^{-1}\right.\right. \\
&\qquad \left.\left. + 2\beta\left(\frac{E_b}{N_0}\right)^{-2} - \frac{8\Phi}{\beta}\sqrt{\frac{P_B}{P_s}}\right]^{-\frac{1}{2}}\right) \\
&+ \frac{1}{4}\mathrm{erfc}\left(\left(1 - \frac{P_B}{P_s}\right)\left[\frac{2\Psi}{\beta} + \frac{4P_B}{P_s}\left(\frac{E_b}{N_0}\right)^{-1}\right.\right.
\end{aligned}$$

$$+4\left(\frac{E_b}{N_0}\right)^{-1}+2\beta\left(\frac{E_b}{N_0}\right)^{-2}\Bigg]^{-\frac{1}{2}}\Bigg) \tag{7.99}$$

where Ψ and Φ are constants as defined in (7.43) and (7.44), and $E_b = 2\beta P_s$ denotes the average bit energy. The expression given in (7.98) or (7.99) is the analytical BER for the noisy DCSK signal in a combined communication environment. Note that for fixed BPSK signal power P_B and noise power spectral density $N_0/2$, the BER can be reduced by making one or a combination of the following adjustments.

1. Reduce the variance of $\{x_k^2\}$ for a given P_s.
2. Reduce the absolute value of Ω, i.e., $\mathrm{E}[x_{k+1}x_k^2]$, for a given P_s.
3. Increase the spreading factor 2β.
4. Increase the DCSK signal power P_s.

In particular, when the BPSK signal power is zero, i.e., $P_B = 0$, it can be readily shown that the bit error rate reduces to (see Sect. 5.2.2)

$$\mathrm{BER}_{\mathrm{DCSK}} \mid (P_B = 0) = \frac{1}{2}\mathrm{erfc}\left[\left(\frac{2\Psi}{\beta}+\frac{2N_0}{\beta P_s}+\frac{N_0^2}{2\beta P_s^2}\right)^{-\frac{1}{2}}\right] \tag{7.100}$$

$$= \frac{1}{2}\mathrm{erfc}\left[\left(\frac{2\Psi}{\beta}+4\left(\frac{E_b}{N_0}\right)^{-1}+2\beta\left(\frac{E_b}{N_0}\right)^{-2}\right)^{-\frac{1}{2}}\right]. \tag{7.101}$$

7.4.2 Performance of the BPSK System in Combined DCSK-BPSK System

The same BPSK receiver shown in Fig. 7.4 is used to demodulate the BPSK signal in the combined DCSK-BPSK communication system. The output of the summer at the end of the lth symbol duration is

$$\begin{aligned} z_l &= \sum_{k=2\beta(l-1)+1}^{2\beta l} r_k \\ &= \underbrace{(1+\alpha_l)\sum_{k=2\beta(l-1)+1}^{2\beta(l-1)+\beta} x_k}_{\text{interfering DCSK signal}} + \underbrace{2\beta\sqrt{P_B}b_l}_{\text{required signal}} \\ &\quad + \underbrace{\sum_{k=2\beta(l-1)+1}^{2\beta l} \xi_k}_{\text{noise}} \,. \end{aligned} \tag{7.102}$$

When the transmitted symbol for the DCSK system is "-1", i.e., $\alpha_l = -1$, (7.102) becomes

$$z_l \mid (\alpha_l = -1) = 2\beta\sqrt{P_B}b_l + \sum_{k=2\beta(l-1)+1}^{2\beta l} \xi_k. \tag{7.103}$$

Clearly, from (7.103), the interference from the DCSK signal vanishes. This is because the interfering DCSK signals coming from the first half and second half of the symbol duration exactly cancel each other. When $\alpha_l = +1$, (7.102) becomes

$$\begin{aligned} z_l \mid (\alpha_l = +1) &= 2\sum_{k=2\beta(l-1)+1}^{2\beta(l-1)+\beta} x_k + 2\beta\sqrt{P_B}b_l \\ &+ \sum_{k=2\beta(l-1)+1}^{2\beta l} \xi_k. \end{aligned} \tag{7.104}$$

Using a likewise procedure as in Sect. 7.3.2, it can be shown that the means and variances of $z_l \mid (\alpha_l = -1)$ and $z_l \mid (\alpha_l = +1)$, denoted by $\mathrm{E}[z_l \mid (\alpha_l = -1)]$, $\mathrm{E}[z_l \mid (\alpha_l = +1)]$, $\mathrm{var}[z_l \mid (\alpha_l = -1)]$ and $\mathrm{var}[z_l \mid (\alpha_l = +1)]$, respectively, are given by

$$\mathrm{E}[z_l \mid (\alpha_l = -1)] = 2\beta\sqrt{P_B}b_l \tag{7.105}$$

$$\mathrm{E}[z_l \mid (\alpha_l = +1)] = 2\mathrm{E}\left[\sum_{k=2\beta(l-1)+1}^{2\beta(l-1)+\beta} x_k\right] + 2\beta\sqrt{P_B}b_l \tag{7.106}$$

$$\mathrm{var}[z_l \mid (\alpha_l = -1)] = \beta N_0 \tag{7.107}$$

$$\mathrm{var}[z_l \mid (\alpha_l = +1)] = 4\mathrm{var}\left[\sum_{k=2\beta(l-1)+1}^{2\beta(l-1)+\beta} x_k\right] + \beta N_0. \tag{7.108}$$

Assuming that (i) the mean value of $\{x_k\}$ is zero, and (ii) the autovariance of $\{x_k\}$ vanishes, we combine (7.27) and (7.28) with (7.106) and (7.108) to get

$$\mathrm{E}[z_l \mid (\alpha_l = +1)] = 2\beta\sqrt{P_B}b_l \tag{7.109}$$

$$\mathrm{var}[z_l \mid (\alpha_l = +1)] = 4\beta P_s + \beta N_0. \tag{7.110}$$

Suppose $b_l = +1$ and $\alpha_l = +1$. As $z_l \mid (b_l = +1, \alpha_l = +1)$ is the sum of a large number of random variables, we may assume that it follows a normal distribution. An error occurs when $z_l \leq 0 \mid (b_l = +1, \alpha_l = +1)$, and the corresponding error probability is given by

$$\begin{aligned} &\mathrm{Prob}(z_l \leq 0 \mid (b_l = +1, \alpha_l = +1)) \\ &= \frac{1}{2}\mathrm{erfc}\left[\frac{\mathrm{E}\left[z_l \mid (b_l = +1, \alpha_l = +1)\right]}{\sqrt{2\mathrm{var}\left[z_l \mid (b_l = +1, \alpha_l = +1)\right]}}\right] \end{aligned}$$

$$= \frac{1}{2}\text{erfc}\left[\frac{2\beta\sqrt{P_B}}{\sqrt{8\beta P_s + 2\beta N_0}}\right]. \quad (7.111)$$

Likewise, it can be shown that

$$\text{Prob}(z_l \le 0 \mid (b_l = +1, \alpha_l = -1)) = \frac{1}{2}\text{erfc}\left[\frac{2\beta\sqrt{P_B}}{\sqrt{2\beta N_0}}\right] \quad (7.112)$$

$$\text{Prob}(z_l > 0 \mid (b_l = -1, \alpha_l = +1)) = \frac{1}{2}\text{erfc}\left[\frac{2\beta\sqrt{P_B}}{\sqrt{8\beta P_s + 2\beta N_0}}\right] \quad (7.113)$$

$$\text{Prob}(z_l > 0 \mid (b_l = -1, \alpha_l = -1)) = \frac{1}{2}\text{erfc}\left[\frac{2\beta\sqrt{P_B}}{\sqrt{2\beta N_0}}\right]. \quad (7.114)$$

The error probability for the lth transmitted BPSK symbol is given by

$$\begin{aligned}
\text{BER}^{(l)}_{\text{BPSK-DCSK}} &= \text{Prob}(b_l = +1) \times \text{Prob}(\alpha_l = +1) \\
&\quad \times \text{Prob}(z_l \le 0 \mid (b_l = +1, \alpha_l = +1)) \\
&\quad + \text{Prob}(b_l = +1) \times \text{Prob}(\alpha_l = -1) \\
&\quad \times \text{Prob}(z_l \le 0 \mid (b_l = +1, \alpha_l = -1)) \\
&\quad + \text{Prob}(b_l = -1) \times \text{Prob}(\alpha_l = +1) \\
&\quad \times \text{Prob}(z_l > 0 \mid (b_l = -1, \alpha_l = +1)) \\
&\quad + \text{Prob}(b_l = -1) \times \text{Prob}(\alpha_l = -1) \\
&\quad \times \text{Prob}(z_l > 0 \mid (b_l = -1, \alpha_l = -1)) \\
&= \frac{1}{4}\text{erfc}\left(\frac{2\beta\sqrt{P_B}}{\sqrt{8\beta P_s + 2\beta N_0}}\right) + \frac{1}{4}\text{erfc}\left(\frac{2\beta\sqrt{P_B}}{\sqrt{2\beta N_0}}\right). \quad (7.115)
\end{aligned}$$

Since $\text{BER}^{(l)}_{\text{BPSK-DCSK}}$ is independent of l, the error probability of the lth transmitted symbol is the same as the BER of the system. Therefore, the BER of the BPSK system, denoted by $\text{BER}_{\text{BPSK-DCSK}}$, is

$$\begin{aligned}
\text{BER}_{\text{BPSK-DCSK}} &= \text{BER}^{(l)}_{\text{BPSK-DCSK}} \\
&= \frac{1}{4}\text{erfc}\left(\frac{2\beta\sqrt{P_B}}{\sqrt{8\beta P_s + 2\beta N_0}}\right) + \frac{1}{4}\text{erfc}\left(\frac{2\beta\sqrt{P_B}}{\sqrt{2\beta N_0}}\right) \quad (7.116) \\
&= \frac{1}{4}\text{erfc}\left(\sqrt{\frac{E_B}{\breve{N}_0}}\right) + \frac{1}{4}\text{erfc}\left(\sqrt{\frac{E_B}{N_0}}\right) \quad (7.117)
\end{aligned}$$

where E_B is as defined in (7.57) and

$$\frac{\breve{N}_0}{2} = \frac{N_0}{2} + 2P_s \quad (7.118)$$

represents the equivalent noise power spectral density when the interfering DCSK signal is taken into consideration. It can be seen that the BPSK signal remains unaffected by the DCSK signal for half of the time and the noise power affecting the BPSK signal increases by $2P_s$ for another half of the time. The expression given in (7.116) or (7.117) is the analytical BER for the noisy coherent BPSK system in a combined communication environment. For a fixed chaotic signal power P_s, the BER can be improved by increasing the spreading factor 2β and/or increasing the BPSK signal power P_B.

Example — Consider the case where the logistic map described in Sect. 7.3.2 is used for generating the chaotic sequences. We substitute (7.61) to (7.63) into (7.98) to obtain the BER of the DCSK system, i.e.,

$$\begin{aligned}
&\mathrm{BER}_{\mathrm{DCSK}} \\
&\approx \frac{1}{8}\mathrm{erfc}\left(\frac{0.5\beta+\beta P_B}{\sqrt{0.25\beta+4\beta P_B+2\beta P_B N_0+\beta N_0+\frac{\beta N_0^2}{2}-2\beta\sqrt{P_B}}}\right) \\
&+ \frac{1}{8}\mathrm{erfc}\left(\frac{0.5\beta+\beta P_B}{\sqrt{0.25\beta+4\beta P_B+2\beta P_B N_0+\beta N_0+\frac{\beta N_0^2}{2}+2\beta\sqrt{P_B}}}\right) \\
&+ \frac{1}{4}\mathrm{erfc}\left(\frac{0.5\beta-\beta P_B}{\sqrt{0.25\beta+2\beta P_B N_0+\beta N_0+\frac{\beta N_0^2}{2}}}\right). \qquad (7.119)
\end{aligned}$$

For the BPSK system, we combine (7.61) with (7.116) to obtain

$$\mathrm{BER}_{\mathrm{BPSK\text{-}DCSK}} = \frac{1}{4}\mathrm{erfc}\left(\sqrt{\frac{2\beta P_B}{2+N_0}}\right)+\frac{1}{4}\mathrm{erfc}\left(\sqrt{\frac{2\beta P_B}{N_0}}\right). \qquad (7.120)$$

7.5 Simulations and Evaluation

In this section we study the performances of the chaos-based and conventional digital communication systems under a combined environment by computer simulations. The logistic map is used to generate the chaotic sequences. In particular, the BER performance of each of the chaos-based and conventional communication systems will be investigated under variation of the following parameters:

- Average bit-energy-to-noise-spectral-density ratio (E_b/N_0 and E_B/N_0)
- Conventional-to-chaotic-signal-power ratio
- Spreading factor

For comparison, we also plot in each case the analytical BERs obtained from the expressions derived in Sects. 7.3 and 7.4. Results are shown in Figs. 7.6 and 7.7 for the combined CSK-BPSK system, and in Figs. 7.8 and 7.9 for the combined DCSK-BPSK system. In general, computer simulations and analytical results are in good agreement. Also, as would be expected, the coherent CSK system generally performs better than the non-coherent DCSK system. Further observations are summarized as follows.

- Except for the DCSK system, the BERs of the combined chaos-based and conventional systems generally decreases (improves) as the spreading factor or the bit-energy-to-noise-power-spectral-density (E_b/N_0 or E_B/N_0) increases.
- The BER of the chaos-based system in the combined environment generally deteriorates (increases) as P_B/P_s increases for any given E_b/N_0. This is apparently due to the increasing power of the BPSK signal which causes more interference to the chaos-based system, thus giving a higher BER.
- At a fixed E_B/N_0, the BER of the BPSK system in the combined environment improves as P_B/P_s increases. This result comes with no surprise because as P_B/P_s increases, the power of the chaotic signal becomes weaker compared to the BPSK signal power. Thus, the interference due to the chaotic signal diminishes, resulting in an improved BER for the BPSK system.
- Comparing the two types of chaos-based communication systems, the performance of the DCSK system is degraded to a larger extent under the influence of a BPSK signal. For example, from Fig. 7.8, for a spreading factor of 200 and $E_b/N_0 = 20$ dB, we observe that the BER of the DCSK system increases from 2×10^{-4} to 0.5 when P_B/P_s increases from -5 dB to 5 dB. For the CSK system employing the same spreading factor, at $E_b/N_0 = 8$ dB, the BER only increases from around 2×10^{-4} to 7×10^{-4} when P_B/P_s increases from -5 dB to 5 dB.
- From Figs. 7.6 and 7.7, we observe that both the BPSK and CSK systems can perform reasonably well under a combined environment. Moreover, from Figs. 7.8 and 7.9, we see that the DCSK-BPSK system generally requires more restrictive operating conditions in order to maintain performance.

Finally, it is of interest to examine the channel capacity for a given total-bit-energy-to-noise-power-spectral-density ratio, defined as E_t/N_0, where $E_t = E_b + E_B$. The total capacity of the combined system is the sum of the capacities of the constituent systems. The capacity of each constituent system can be evaluated using the capacity formula for a binary symmetric channel. Thus, the total capacity, C_{cap}, for the CSK-BPSK system is given by

$$C_{\text{cap}} = (1 - H(\text{BER}_{\text{CSK}})) + (1 - H(\text{BER}_{\text{BPSK-CSK}})) \tag{7.121}$$

and that for the DCSK-BPSK system is

$$C_{\text{cap}} = (1 - H(\text{BER}_{\text{DCSK}})) + (1 - H(\text{BER}_{\text{BPSK-DCSK}})), \tag{7.122}$$

where $H(.)$ represents the entropy function [Proakis and Salehi (1994)]. The results are plotted in Fig. 7.10. A few observations concerning the system capacities can be made here:

- Under the same condition, the capacity of the combined CSK-BPSK system is higher than that of the combined DCSK-BPSK system. This is apparently due to better bit error performance of the CSK-BPSK system.

- For the combined CSK-BPSK system, under the same E_t/N_0, the capacity is highest when $P_B/P_s = 0$ dB, i.e., both the CSK and BPSK signals have the same (average) power. This result comes at no surprise because, from the assumptions made on the statistics of the chaotic signal, both the CSK and the BPSK systems should have similar BER performance when their powers are equal. For a fixed power ratio between the chaos and conventional signals, the capacity is the same regardless of which signal having a higher power. In other words, the capacity is the same for the same absolute value of P_B/P_s in dB. Moreover, when the power ratio increases, the capacity decreases.
- For the combined DCSK-BPSK system, the capacity is very low when E_t/N_0 is less than 8 dB. Under the same E_t/N_0 with value above 8 dB, the highest capacity is achieved when $P_B/P_s = -5$ dB.

As a final remark, the afore-described analysis has assumed a synchronous operation of the chaos-based (CSK or DCSK) communication system and the conventional (BPSK) system. We also assume identical bit rates of the coexisting systems. However, in general, the systems may not be synchronized and they may operate at different data rates. Thus, the performances of the real systems may deviate from the reported results. Nonetheless, our results provide a benchmark indicator of the performance of the coexisting systems. In the following chapter, we will continue our study of the coexistence problem and will extend our analysis to include conventional wideband systems.

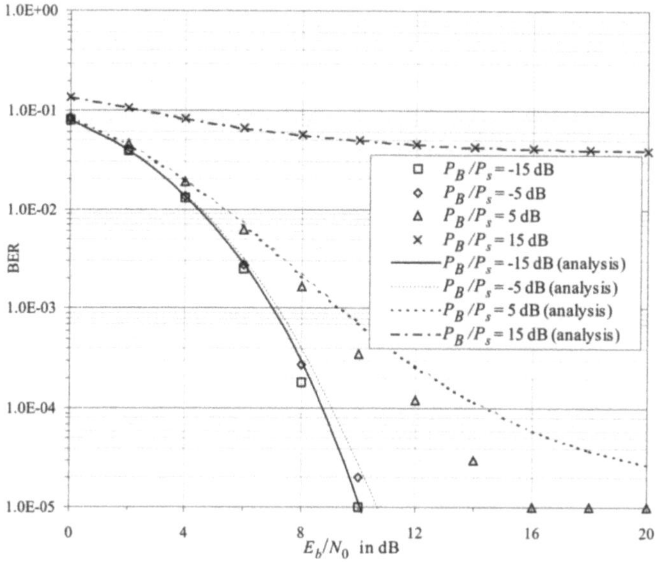

(a)

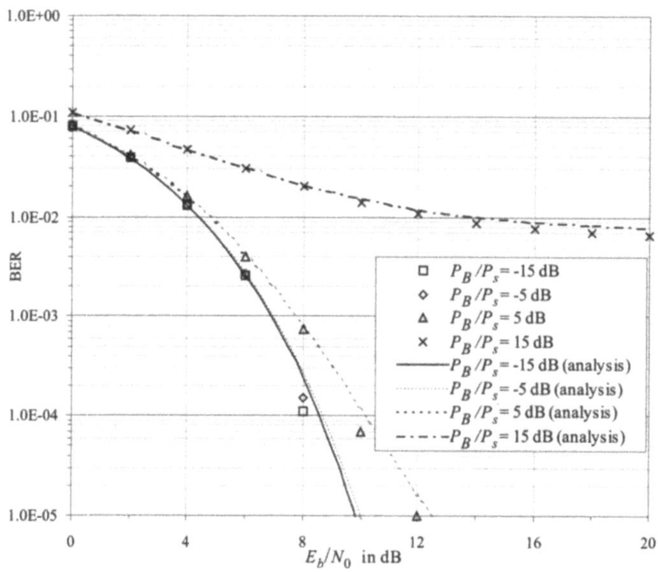

(b)

Fig. 7.6. BERs versus E_b/N_0 of the coherent CSK system in a combined CSK-BPSK environment. Simulated BERs are plotted as points and analytical BERs plotted as lines. (a) Spreading factor is 100; (b) spreading factor is 200.

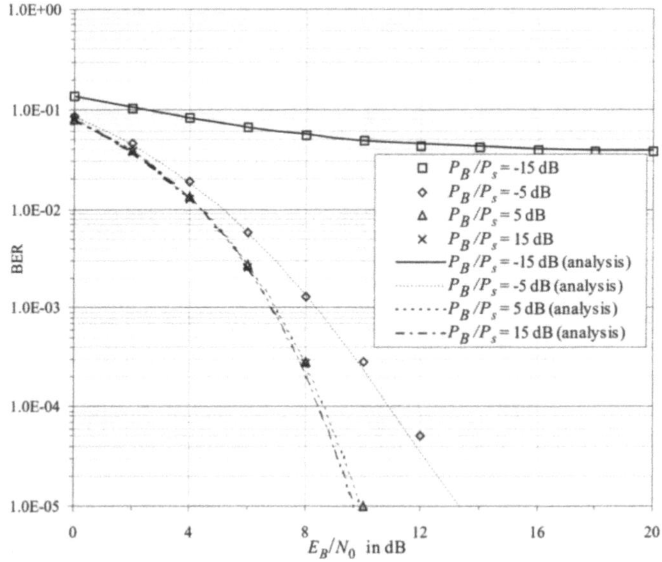

(a)

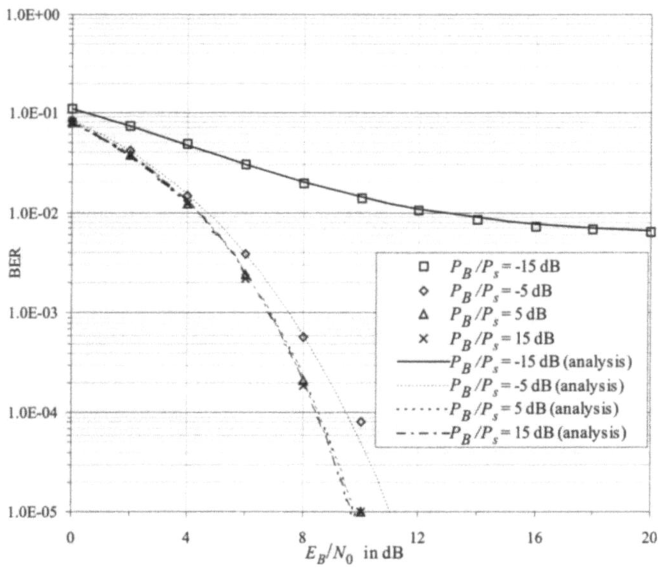

(b)

Fig. 7.7. BERs versus E_B/N_0 of the BPSK system in a combined CSK-BPSK environment. Simulated BERs are plotted as points and analytical BERs plotted as lines. (a) Spreading factor is 100; (b) spreading factor is 200.

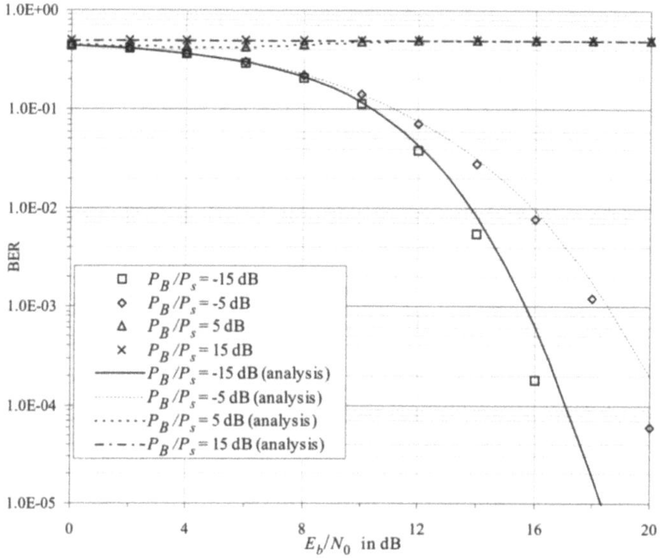

(a)

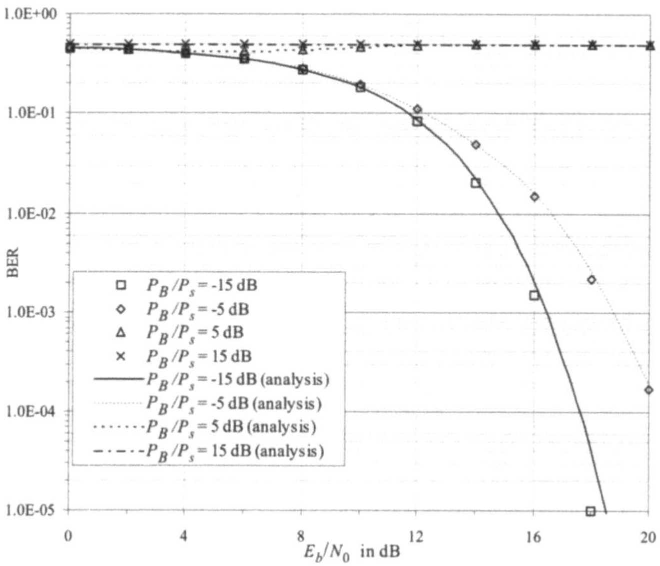

(b)

Fig. 7.8. BERs versus E_b/N_0 of the non-coherent DCSK system in a combined DCSK-BPSK environment. Simulated BERs are plotted as points and analytical BERs plotted as lines. (a) Spreading factor is 100; (b) spreading factor is 200.

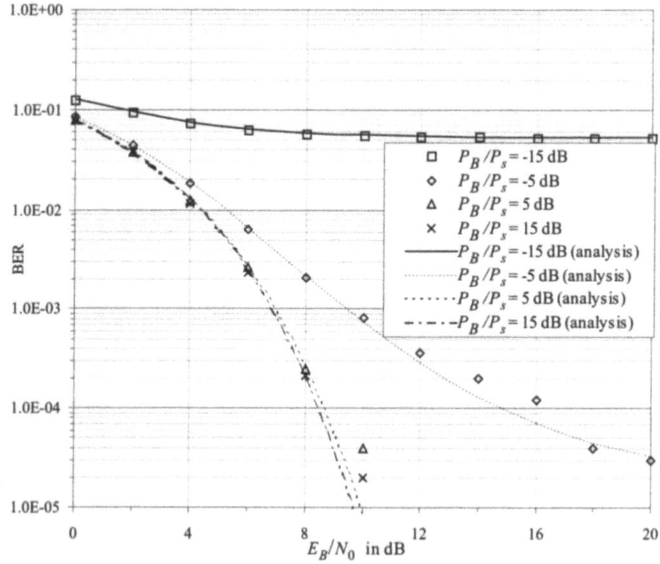

(a)

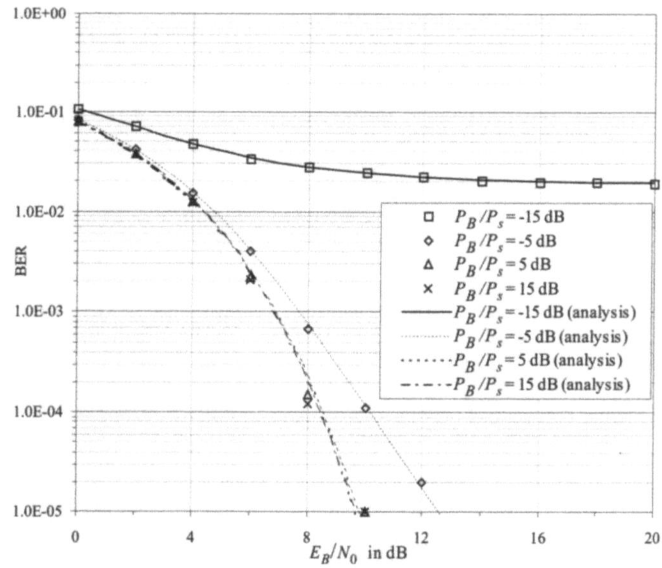

(b)

Fig. 7.9. BERs versus E_B/N_0 of the BPSK system in a combined DCSK-BPSK environment. Simulated BERs are plotted as points and analytical BERs plotted as lines. (a) Spreading factor is 100; (b) spreading factor is 200.

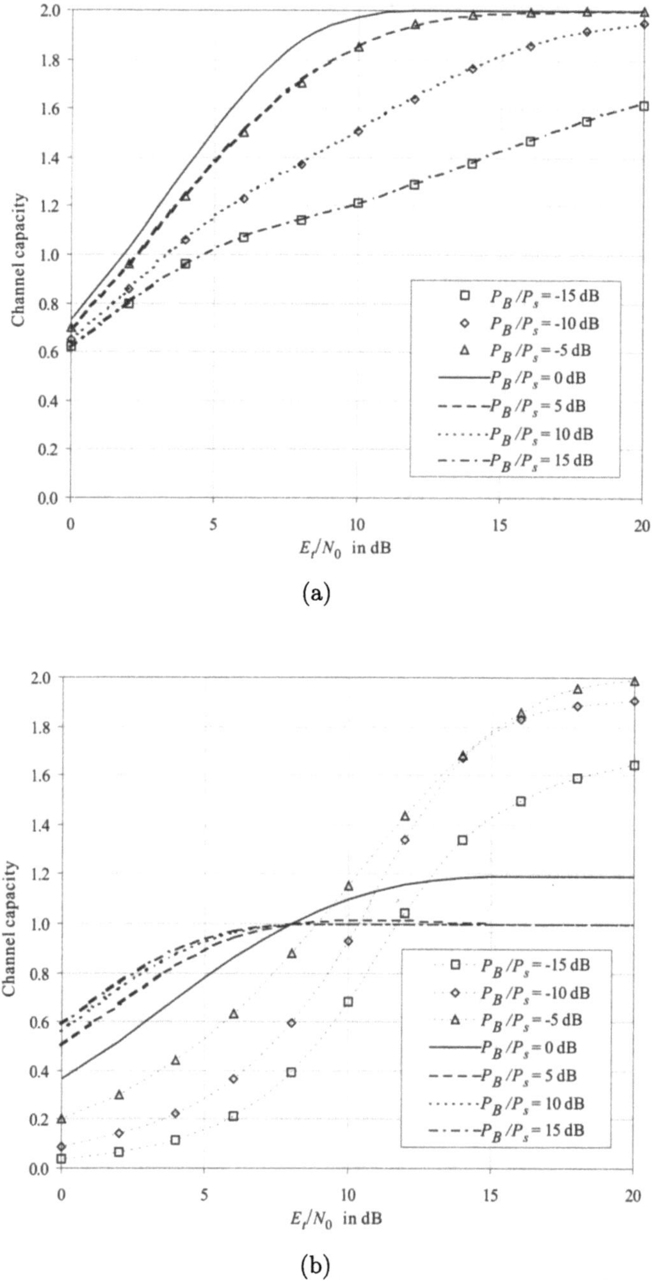

Fig. 7.10. Channel capacity of the combined chaos-based-conventional system. Spreading factor is 100. (a) Combined CSK-BPSK system; (b) combined DCSK-BPSK system.

Appendix 7A: Derivation of covariances and variances relevant to the analysis of combined CSK-BPSK system

All symbols are as defined in Sect. 7.3.

$$
\begin{aligned}
\mathrm{cov}[A,C] &= \mathrm{E}[AC]-\mathrm{E}[A]\mathrm{E}[C]\\
&= \mathrm{E}\left[\left(\sum_{k=2\beta(l-1)+1}^{2\beta l} x_k^2\right)\left(\sum_{k=2\beta(l-1)+1}^{2\beta l} \xi_k x_k\right)\right]\\
&\quad -\mathrm{E}\left[\sum_{k=2\beta(l-1)+1}^{2\beta l} x_k^2\right]\mathrm{E}\left[\sum_{k=2\beta(l-1)+1}^{2\beta l} \xi_k x_k\right]\\
&= \mathrm{E}\left[\sum_{k=2\beta(l-1)+1}^{2\beta l} x_k^3\xi_k+\sum_{\substack{k=2\beta(l-1)+1\\k\neq m}}^{2\beta l}\sum_{m=2\beta(l-1)+1}^{2\beta l} \xi_k x_k x_m^2\right]\\
&\quad -\left(\sum_{k=2\beta(l-1)+1}^{2\beta l}\mathrm{E}[x_k^2]\right)\left(\sum_{k=2\beta(l-1)+1}^{2\beta l}\mathrm{E}[\xi_k]\mathrm{E}[x_k]\right)\\
&= \sum_{k=2\beta(l-1)+1}^{2\beta l}\mathrm{E}[x_k^3]\mathrm{E}[\xi_k]\\
&\quad +\sum_{\substack{k=2\beta(l-1)+1\\k\neq m}}^{2\beta l}\sum_{m=2\beta(l-1)+1}^{2\beta l}\mathrm{E}[\xi_k]\mathrm{E}[x_k x_m^2]\\
&= 0
\end{aligned}
\tag{7.123}
$$

$$
\begin{aligned}
\mathrm{cov}[B,C] &= \mathrm{E}[BC]-\mathrm{E}[B]\mathrm{E}[C]\\
&= \mathrm{E}\left[\left(\sqrt{P_B}\sum_{k=2\beta(l-1)+1}^{2\beta l} x_k\right)\left(\sum_{k=2\beta(l-1)+1}^{2\beta l} \xi_k x_k\right)\right]\\
&\quad -\mathrm{E}\left[\sqrt{P_B}\sum_{k=2\beta(l-1)+1}^{2\beta l} x_k\right]\mathrm{E}\left[\sum_{k=2\beta(l-1)+1}^{2\beta l} \xi_k x_k\right]\\
&= \sqrt{P_B}\mathrm{E}\left[\sum_{k=2\beta(l-1)+1}^{2\beta l} x_k^2\xi_k+\sum_{\substack{k=2\beta(l-1)+1\\k\neq m}}^{2\beta l}\sum_{m=2\beta(l-1)+1}^{2\beta l} \xi_k x_k x_m\right]\\
&\quad -\sqrt{P_B}\left(\sum_{k=2\beta(l-1)+1}^{2\beta l}\mathrm{E}[x_k]\right)\left(\sum_{k=2\beta(l-1)+1}^{2\beta l}\mathrm{E}[\xi_k]\mathrm{E}[x_k]\right)
\end{aligned}
$$

$$
\begin{aligned}
&= \sqrt{P_B}\left[\sum_{k=2\beta(l-1)+1}^{2\beta l} \mathrm{E}[x_k^2]\mathrm{E}[\xi_k] + \sum_{\substack{k=2\beta(l-1)+1\\ k\neq m}}^{2\beta l}\sum_{m=2\beta(l-1)+1}^{2\beta l} \mathrm{E}[\xi_k]\mathrm{E}[x_k x_m]\right] \\
&= 0 \qquad (7.124)
\end{aligned}
$$

$$
\begin{aligned}
\mathrm{cov}[A,B] &= \mathrm{cov}\left[\sum_{k=2\beta(l-1)+1}^{2\beta l} x_k^2, \sqrt{P_B}\sum_{k=2\beta(l-1)+1}^{2\beta l} x_k\right] \\
&= \sqrt{P_B}\mathrm{E}\left[\left(\sum_{k=2\beta(l-1)+1}^{2\beta l} x_k^2\right)\left(\sum_{k=2\beta(l-1)+1}^{2\beta l} x_k\right)\right] - \sqrt{P_B}\mathrm{E}\left[\sum_{k=2\beta(l-1)+1}^{2\beta l} x_k^2\right]\mathrm{E}\left[\sum_{k=2\beta(l-1)+1}^{2\beta l} x_k\right] \\
&= \sqrt{P_B}\mathrm{E}\left[\sum_{k=2\beta(l-1)+1}^{2\beta l}\sum_{m=2\beta(l-1)+1}^{2\beta l} x_k x_m^2\right] - \sqrt{P_B}\left(\sum_{k=2\beta(l-1)+1}^{2\beta l}\mathrm{E}[x_k^2]\right)\left(\sum_{k=2\beta(l-1)+1}^{2\beta l}\mathrm{E}[x_k]\right) \\
&= \sqrt{P_B}\sum_{k=2\beta(l-1)+1}^{2\beta l}\sum_{m=2\beta(l-1)+1}^{2\beta l}\mathrm{E}[x_k x_m^2] - \sqrt{P_B}\left(\sum_{k=2\beta(l-1)+1}^{2\beta l}\mathrm{E}[x_k^2]\right)\left(\sum_{k=2\beta(l-1)+1}^{2\beta l}\mathrm{E}[x_k]\right) \\
&= \sqrt{P_B}\sum_{k=2\beta(l-1)+1}^{2\beta l-1}\mathrm{E}[x_{k+1}x_k^2] \qquad \text{apply (7.27) and (7.30)} \\
&= \sqrt{P_B}(2\beta-1)\mathrm{E}[x_{k+1}x_k^2] \qquad (7.125)
\end{aligned}
$$

$$
\begin{aligned}
\mathrm{var}[A] &= \mathrm{var}\left[\sum_{k=2\beta(l-1)+1}^{2\beta l} x_k^2\right] \\
&= \sum_{k=2\beta(l-1)+1}^{2\beta l}\mathrm{var}[x_k^2] + \sum_{\substack{k=2\beta(l-1)+1\\ k\neq m}}^{2\beta l}\sum_{m=2\beta(l-1)+1}^{2\beta l}\mathrm{cov}[x_k^2, x_m^2] \\
&= 2\beta\mathrm{var}[x_k^2] \qquad \text{apply (7.29)} \\
&= 2\beta\Lambda \qquad \text{where } \Lambda = \mathrm{var}[x_k^2] \qquad (7.126)
\end{aligned}
$$

$$
\begin{aligned}
\mathrm{var}[B] &= \mathrm{var}\left[\sqrt{P_B}\sum_{k=2\beta(l-1)+1}^{2\beta l} x_k\right] \\
&= P_B \sum_{k=2\beta(l-1)+1}^{2\beta l} \mathrm{var}[x_k] \\
&\quad + P_B \sum_{\substack{k=2\beta(l-1)+1 \\ k\neq m}}^{2\beta l} \sum_{m=2\beta(l-1)+1}^{2\beta l} \mathrm{cov}[x_k, x_m] \\
&= 2\beta P_B P_s \qquad \text{apply (7.28)}
\end{aligned} \tag{7.127}
$$

$$
\begin{aligned}
\mathrm{var}[C] &= \mathrm{var}\left[\sum_{k=2\beta(l-1)+1}^{2\beta l} \xi_k x_k\right] \\
&= \sum_{k=2\beta(l-1)+1}^{2\beta l} \mathrm{var}[\xi_k x_k] \\
&\quad + \sum_{\substack{k=2\beta(l-1)+1 \\ k\neq m}}^{2\beta l} \sum_{m=2\beta(l-1)+1}^{2\beta l} \mathrm{cov}[\xi_k x_k, \xi_m x_m] \\
&= \sum_{k=2\beta(l-1)+1}^{2\beta l} \left\{\mathrm{E}[\xi_k^2 x_k^2] - \mathrm{E}^2[\xi_k x_k]\right\} \\
&\quad + \sum_{\substack{k=2\beta(l-1)+1 \\ k\neq m}}^{2\beta l} \sum_{m=2\beta(l-1)+1}^{2\beta l} \left\{\mathrm{E}[\xi_k x_k \xi_m x_m] - \mathrm{E}[\xi_k x_k]\mathrm{E}[\xi_m x_m]\right\} \\
&= \sum_{k=2\beta(l-1)+1}^{2\beta l} \left\{\mathrm{E}[\xi_k^2]\mathrm{E}[x_k^2] - \mathrm{E}^2[\xi_k]\mathrm{E}^2[x_k]\right\} \\
&\quad + \sum_{\substack{k=2\beta(l-1)+1 \\ k\neq m}}^{2\beta l} \sum_{m=2\beta(l-1)+1}^{2\beta l} \left\{\mathrm{E}[x_k x_m]\mathrm{E}[\xi_k \xi_m] \right. \\
&\qquad\qquad \left. - \mathrm{E}[x_k]\mathrm{E}[x_m]\mathrm{E}[\xi_k]\mathrm{E}[\xi_m]\right\} \\
&= 2\beta\left(\frac{N_0}{2}\right)P_s \\
&= \beta N_0 P_s
\end{aligned} \tag{7.128}
$$

Appendix 7B: Derivation of $\mathrm{E}[x_k x_m^2]$ for chaotic sequences generated by the logistic map

All symbols are as defined in Sect. 7.3. For $k = m$,

$$\mathrm{E}[x_k x_m^2] = \mathrm{E}[x_k^3] = \int_{-\infty}^{\infty} x^3 \rho(x) dx = \int_{-1}^{1} x^3 \rho(x) dx = 0 \tag{7.129}$$

because $\rho(x)$ is an even function whereas x^3 is odd. Next, we consider the case where $k < m$. Assuming $m = k+n$ for some positive integer n, we obtain

$$\mathrm{E}[x_k x_m^2] = \int_{-\infty}^{\infty} \rho(x)\ x\ (g^{(n)}(x))^2 dx. \tag{7.130}$$

Within the integral, both $\rho(x)$ and $(g^{(n)}(x))^2$ are even while x is odd. Thus, it can be concluded that $\mathrm{E}[x_k x_m^2] = 0$ for $k < m$. Finally, for $k > m$, we assume $k = m + p$ for some positive integer p. Making the substitution $x = \cos\phi$ and applying (3.61), we have

$$\begin{aligned} \mathrm{E}[x_{m+p} x_m^2] &= \int_{-\infty}^{\infty} \rho(x)\ g^{(p)}(x)\ x^2\ dx \\ &= \frac{1}{\pi} \int_0^{\pi} (-\cos(2^p \phi))\ \cos^2\phi\ d\phi \\ &= \begin{cases} 0 & \text{for } p \geq 2 \\ -\frac{1}{4} & \text{for } p = 1. \end{cases} \end{aligned} \tag{7.131}$$

Thus, we conclude that

$$\mathrm{E}[x_k x_m^2] = \begin{cases} 0 & \text{for } k - m \neq 1 \\ -\frac{1}{4} & \text{for } k - m = 1. \end{cases} \tag{7.132}$$

8. Coexistence of Chaos-Based and Conventional Spread-Spectrum Systems

In this chapter we continue our investigation of the coexistence of chaos-based communication systems and conventional systems. As mentioned in the previous chapter, our motivation in performing this analysis is the possible co-operation of chaos-based and conventional systems under mutual interference. Specifically, being spread-spectrum, chaos-based communication systems are expected to perform reasonably well even in the presence of other wideband signals sharing the same bandwidth. Such a scenario may occur in normal practice, for example, when chaos-based systems are introduced while the conventional systems are still in operation. This aspect of performance is important, though it has rarely been addressed in the literature. It is therefore of interest to probe into the error performance of chaos-based systems in channels where other wideband communication systems coexist. Furthermore, it is useful to compare the relative performances of coherent and non-coherent chaos-based systems and the extent to which coherent systems excel in the presence of other coexisting wideband systems. Our objective in this chapter is to investigate the performance of selected chaos-based digital systems when their bandwidths are co-occupied by a conventional spread-spectrum signal. The chaos-based systems under study are the coherent CSK and the non-coherent DCSK systems, and the coexisting system is a standard direct-sequence spread-spectrum (DS-SS) system [Proakis and Salehi (1994); Ziemer and Tranter (1995)]. Analytical expressions for the bit error rates are derived in terms of system parameters such as spreading factor, chaotic signal power, conventional spread-spectrum signal power and noise power spectral density. Finally, computer simulations are performed to verify the analytical findings.

8.1 System Overview

The basic problem we wish to investigate in this chapter is the performance of chaos-based digital communication systems when the channel is under the influence of

– additive white Gaussian noise; and

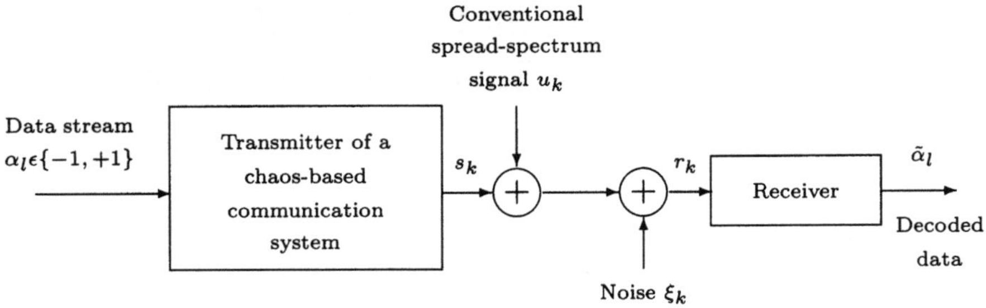

Fig. 8.1. Block diagram of a chaos-based digital communication system with conventional spread-spectrum signal and noise added.

– a wideband signal generated from a coexisting conventional spread-spectrum communication system, which shares the same frequency band as the chaos-based system under study.

Figure 8.1 shows a block diagram of the system under study. Our analysis will be carried out in a discrete-time fashion. At time k, denote the output of the chaos transmitter by s_k and the conventional spread-spectrum signal by u_k. Assuming that noise samples, ξ_k, are added to the transmitted signal samples, the receiver receives signals consisting of three components, namely, chaos, conventional spread-spectrum component and additive noise. This corrupted signal, denoted by r_k, will be used by the receiver to recover the transmitted data stream. Coherent or non-coherent detection schemes may be applied in the receiver, depending upon the modulation method used.

In the following section, we focus on the coherent CSK system and the non-coherent DCSK system [Dedieu *et al.* (1993), Kolumbán *et al.* (1998b)]. The coexisting conventional system is a DS-SS system. We will develop analytical expressions for the bit error rates of the recovered data streams for each of the chaos-based communication schemes under the aforedescribed channel conditions.

8.2 Analysis of Bit Error Performance

8.2.1 Coherent CSK System

We consider a discrete-time binary CSK communication system, as shown in Fig. 8.2. In general, a pair of chaotic sequences, denoted by $\{\hat{x}_k\}$ and $\{\check{x}_k\}$, are generated in the transmitter by two chaotic maps. If the symbol "+1" is sent, $\{\hat{x}_k\}$ is transmitted during a bit period, and if "−1" is sent, $\{\check{x}_k\}$ is transmitted. Further, we assume that "−1" and "+1" occur with equal probabilities. For simplicity, we consider a CSK system in which one chaos generator is used to produce chaotic signal samples $\{x_k\}$ for $k = 1, 2, \ldots$. The two possible transmitted sequences are $\{\hat{x}_k = x_k\}$ and $\{\check{x}_k = -x_k\}$.

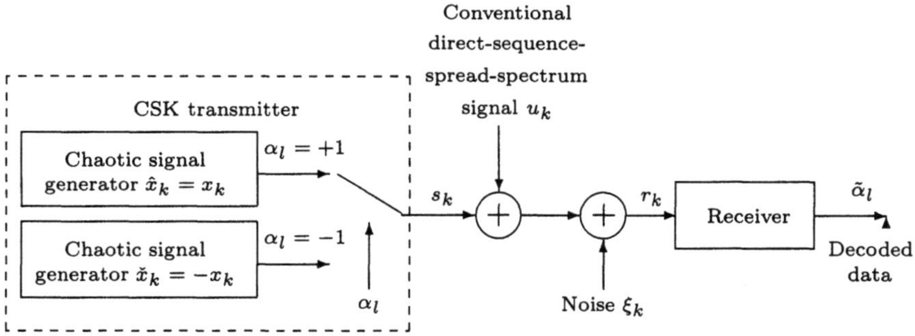

Fig. 8.2. Block diagram of a CSK communication system with conventional direct-sequence-spread-spectrum signal and noise added.

Suppose $\alpha_l \in \{-1, +1\}$ is the symbol to be sent during the lth bit period. Define the spreading factor, 2β, as the number of chaotic samples used to transmit one binary symbol. During the lth bit duration, i.e., for $k = 2\beta(l-1)+1, 2\beta(l-1)+2, \ldots, 2\beta l$, the output of the CSK transmitter is

$$s_k = \alpha_l x_k. \tag{8.1}$$

For the DS-SS system, we assume that the period of the pseudorandom (PN) sequence is very long. We denote the output power of the system by P_B. Essentially, we can model the output of the DS-SS system as a random binary symbol $b_k \in \{-1, +1\}$ multiplied by $\sqrt{P_B}$. Also, "−1" and "+1" occur with equal probabilities. Thus, for time k, the transmitted signal of the DS-SS system is represented by

$$u_k = \sqrt{P_B} b_k. \tag{8.2}$$

The CSK and DS-SS signals are added, and corrupted by an additive white Gaussian noise before arriving at the receiving end. The received signal, denoted by r_k, is thus given by

$$r_k = s_k + \sqrt{P_B} b_k + \xi_k \tag{8.3}$$

where ξ_k is a Gaussian noise sample of zero mean and variance (power spectral density) $N_0/2$. We will first consider the lth transmitted bit and find its error probability, and then derive the error probability for the whole system.

Assume that a correlator-type receiver is employed. As shown in Fig. 8.3, the correlator output for the lth bit, y_l, is given by

$$\begin{aligned} y_l &= \sum_{k=2\beta(l-1)+1}^{2\beta l} r_k x_k \\ &= \underbrace{\alpha_l \sum_{k=2\beta(l-1)+1}^{2\beta l} x_k^2}_{\text{required signal}} + \underbrace{\sqrt{P_B} \sum_{k=2\beta(l-1)+1}^{2\beta l} b_k x_k}_{\text{interfering DS-SS signal}} \end{aligned}$$

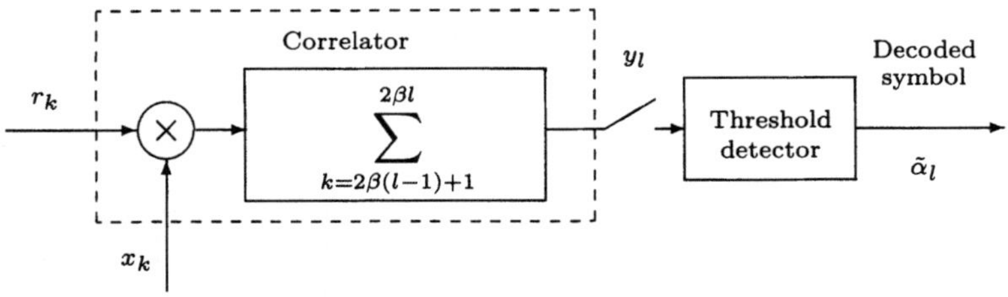

Fig. 8.3. Block diagram of a coherent CSK receiver.

$$+ \underbrace{\sum_{k=2\beta(l-1)+1}^{2\beta l} \xi_k x_k}_{\text{noise}} . \tag{8.4}$$

Suppose a "+1" is transmitted in the CSK system during the lth symbol duration, i.e., $\alpha_l = +1$. For simplicity we write $y_l \mid (\alpha_l = +1)$ as

$$y_l \mid (\alpha_l = +1) = A + B + C \tag{8.5}$$

where A, B and C are the required signal, interfering DS-SS signal and noise, respectively, and are defined as

$$A = \sum_{k=2\beta(l-1)+1}^{2\beta l} x_k^2 \tag{8.6}$$

$$B = \sqrt{P_B} \sum_{k=2\beta(l-1)+1}^{2\beta l} b_k x_k \tag{8.7}$$

$$C = \sum_{k=2\beta(l-1)+1}^{2\beta l} \xi_k x_k . \tag{8.8}$$

The mean of $y_l \mid (\alpha_l = +1)$ is

$$\begin{aligned} \mathrm{E}[y_l \mid (\alpha_l = +1)] &= \mathrm{E}[A] + \mathrm{E}[B] + \mathrm{E}[C] \\ &= \mathrm{E}\left[\sum_{k=2\beta(l-1)+1}^{2\beta l} x_k^2\right] + \mathrm{E}\left[\sqrt{P_B} \sum_{k=2\beta(l-1)+1}^{2\beta l} b_k x_k\right] \\ &\quad + \mathrm{E}\left[\sum_{k=2\beta(l-1)+1}^{2\beta l} \xi_k x_k\right] \\ &= \sum_{k=2\beta(l-1)+1}^{2\beta l} \mathrm{E}[x_k^2] + \sqrt{P_B} \sum_{k=2\beta(l-1)+1}^{2\beta l} \mathrm{E}[b_k]\mathrm{E}[x_k] \end{aligned}$$

$$+ \sum_{k=2\beta(l-1)+1}^{2\beta l} \mathrm{E}[\xi_k]\mathrm{E}[x_k]$$
$$= 2\beta P_s \qquad (8.9)$$

where $P_s = \mathrm{E}[x_k^2]$ denotes the average power of the chaotic signal. The last equality holds because $\mathrm{E}[b_k] = 0$ and $\mathrm{E}[\xi_k] = 0$. The mean value and the average power of the chaotic signal can be computed by numerical simulation, or by numerical integration if the invariant probability density function of $\{x_k\}$ is available. The variance of $y_l \mid (\alpha_l = +1)$ is

$$\begin{aligned} \mathrm{var}[y_l \mid (\alpha_l = +1)] &= \mathrm{var}[A] + \mathrm{var}[B] + \mathrm{var}[C] + 2\mathrm{cov}[A, B] \\ &\quad + 2\mathrm{cov}[B, C] + 2\mathrm{cov}[A, C] \end{aligned} \qquad (8.10)$$

where $\mathrm{cov}[X, Y]$ is the covariance of X and Y defined as

$$\mathrm{cov}[X, Y] = \mathrm{E}[XY] - \mathrm{E}[X]\mathrm{E}[Y]. \qquad (8.11)$$

It can be shown that all the covariance terms are zero and the variance terms are given by (see Appendix 8)

$$\mathrm{var}[A] = 2\beta\Lambda \qquad (8.12)$$
$$\mathrm{var}[B] = 2\beta P_B P_s \qquad (8.13)$$
$$\mathrm{var}[C] = \beta N_0 P_s \qquad (8.14)$$

where Λ is the variance of $\{x_k^2\}$, i.e.,

$$\Lambda = \mathrm{var}[x_k^2]. \qquad (8.15)$$

In the derivation of $\mathrm{var}[A]$, it has been assumed that the autovariance of $\{x_k^2\}$ vanishes, i.e.,

$$\mathrm{cov}[x_k^2, x_m^2] = \mathrm{E}[x_k^2 x_m^2] - \mathrm{E}[x_k^2]\mathrm{E}[x_m^2] = 0 \quad \text{for} \quad k \neq m. \qquad (8.16)$$

Using (8.12) to (8.14), and because all covariance terms in (8.10) are zero, we may write (8.10) as

$$\begin{aligned} \mathrm{var}[y_l \mid (\alpha_l = +1)] &= \mathrm{var}[A] + \mathrm{var}[B] + \mathrm{var}[C] \\ &= 2\beta\Lambda + 2\beta P_B P_s + \beta N_0 P_s. \end{aligned} \qquad (8.17)$$

For the lth symbol, an error occurs when $y_l \leq 0 \mid (\alpha_l = +1)$. Since $y_l \mid (\alpha_l = +1)$ is the sum of a large number of random variables, we may assume that it follows a normal distribution when 2β is large. The error probability is thus given by

$$\mathrm{Prob}(y_l \leq 0 \mid (\alpha_l = +1)) = \frac{1}{2}\mathrm{erfc}\left(\frac{\mathrm{E}\left[y_l \mid (\alpha_l = +1)\right]}{\sqrt{2\mathrm{var}\left[y_l \mid (\alpha_l = +1)\right]}}\right)$$

$$= \frac{1}{2}\text{erfc}\left(\frac{2\beta P_s}{\sqrt{4\beta\Lambda + 4\beta P_B P_s + 2\beta N_0 P_s)}}\right) \tag{8.18}$$

where erfc(.) is the complementary error function, which is defined as

$$\text{erfc}(\psi) \equiv \frac{2}{\sqrt{\pi}}\int_{\psi}^{\infty} e^{-\lambda^2} d\lambda. \tag{8.19}$$

Similarly, when $\alpha_l = -1$, the output of the correlator can be shown equal to

$$y_l \mid (\alpha_l = -1) = -A + B + C. \tag{8.20}$$

Likewise, the mean and variance of $y_l \mid (\alpha_l = -1)$, denoted by $\text{E}[y_l \mid (\alpha_l = -1)]$ and $\text{var}[y_l \mid (\alpha_l = -1)]$, can be derived, i.e.,

$$\text{E}[y_l \mid (\alpha_l = -1)] = -2\beta P_s \tag{8.21}$$

$$\text{var}[y_l \mid (\alpha_l = -1)] = \text{var}[A] + \text{var}[B] + \text{var}[C] \tag{8.22}$$

where A, B and C are defined as in (8.6) to (8.8). The corresponding error probability is

$$\begin{aligned} \text{Prob}(y_l > 0 \mid (\alpha_l = -1)) &= \frac{1}{2}\text{erfc}\left(\frac{-\text{E}\left[y_l \mid (\alpha_l = -1)\right]}{\sqrt{2\text{var}\left[y_l \mid (\alpha_l = -1)\right]}}\right) \\ &= \frac{1}{2}\text{erfc}\left(\frac{2\beta P_s}{\sqrt{4\beta\Lambda + 4\beta P_B P_s + 2\beta N_0 P_s)}}\right). \end{aligned} \tag{8.23}$$

Hence, the overall error probability of the lth transmitted symbol is

$$\begin{aligned} \text{BER}_{\text{CSK}}^{(l)} &= \text{Prob}(\alpha_l = +1) \times \text{Prob}(y_l \le 0 \mid (\alpha_l = +1)) \\ &\quad + \text{Prob}(\alpha_l = -1) \times \text{Prob}(y_l \le 0 \mid (\alpha_l = -1)) \\ &= \frac{1}{2}\text{erfc}\left(\frac{2\beta P_s}{\sqrt{4\beta\Lambda + 4\beta P_B P_s + 2\beta N_0 P_s)}}\right). \end{aligned} \tag{8.24}$$

It can be seen from (8.24) that $\text{BER}_{\text{CSK}}^{(l)}$ is independent of l. Thus, the error probability of the lth transmitted symbol is the same as the bit error rate (BER) of the system. The BER of the CSK system, denoted by BER_{CSK}, is therefore

$$\begin{aligned} \text{BER}_{\text{CSK}} &= \frac{1}{2}\text{erfc}\left(\frac{2\beta P_s}{\sqrt{4\beta\Lambda + 4\beta P_B P_s + 2\beta N_0 P_s)}}\right) \\ &= \frac{1}{2}\text{erfc}\left[\left(\frac{\Psi}{\beta} + \frac{P_B}{\beta P_s} + \frac{N_0}{2\beta P_s}\right)^{-\frac{1}{2}}\right] \qquad (8.25) \\ &= \frac{1}{2}\text{erfc}\left[\left(\frac{\Psi}{\beta} + \frac{P_B}{\beta P_s} + \left(\frac{E_b}{N_0}\right)^{-1}\right)^{-\frac{1}{2}}\right] \qquad (8.26) \end{aligned}$$

where

$$\Psi = \frac{\text{var}[x_k^2]}{P_s^2} \tag{8.27}$$

is a constant depending upon the type of the chaotic sequence (see Sect. 3.2.3.2 for details), and

$$E_b = 2\beta P_s \tag{8.28}$$

denotes the average bit energy of the CSK system. The expression given in (8.26) is thus the analytical BER for the noisy coherent CSK system under the interference of a DS-SS signal. Note that for fixed DS-SS signal power P_B and noise power spectral density $N_0/2$, the BER can be improved by making one or more of the following adjustments.

1. Reduce the variance of $\{x_k^2\}$ for a given P_s.
2. Increase the spreading factor 2β.
3. Increase the CSK signal power P_s.

When the DS-SS signal power is zero, i.e., $P_B = 0$, it can be readily shown that the bit error rate reduces to (see Sect. 5.2.1)

$$\text{BER}_{\text{CSK}} \mid (P_B = 0) = \frac{1}{2}\text{erfc}\left[\left(\frac{\Psi}{\beta} + \frac{N_0}{2\beta P_s}\right)^{-\frac{1}{2}}\right] \tag{8.29}$$

$$= \frac{1}{2}\text{erfc}\left[\left(\frac{\Psi}{\beta} + \left(\frac{E_b}{N_0}\right)^{-1}\right)^{-\frac{1}{2}}\right]. \tag{8.30}$$

Example — Consider the case where the logistic map is used for chaos generation. The form of the map is

$$x_{k+1} = g(x_k) = 1 - 2x_k^2 \text{ where } x_k \in (-1, +1). \tag{8.31}$$

Given that the invariant probability density function of $\{x_k\}$ equals [Kohda and Tsuneda (1994)]

$$\rho(x_k) = \begin{cases} \dfrac{1}{\pi\sqrt{1 - x_k^2}} & \text{if } |x_k| < 1 \\ 0 & \text{otherwise,} \end{cases} \tag{8.32}$$

we obtain

$$P_s = \text{E}[x_k^2] = \int_{-\infty}^{\infty} x^2 \rho(x) dx = \int_{-1}^{1} x^2 \rho(x) dx = \frac{1}{2} \tag{8.33}$$

$$\Lambda = \text{var}[x_k^2] = \text{E}[x_k^4] - \text{E}^2[x_k^2] = \int_{-1}^{1} x^4 \rho(x) dx - \frac{1}{4} = \frac{1}{8}. \tag{8.34}$$

For the case where the logistic map is used to generate the chaotic samples, it can be shown that the assumption given in (8.16) is valid (see Appendix

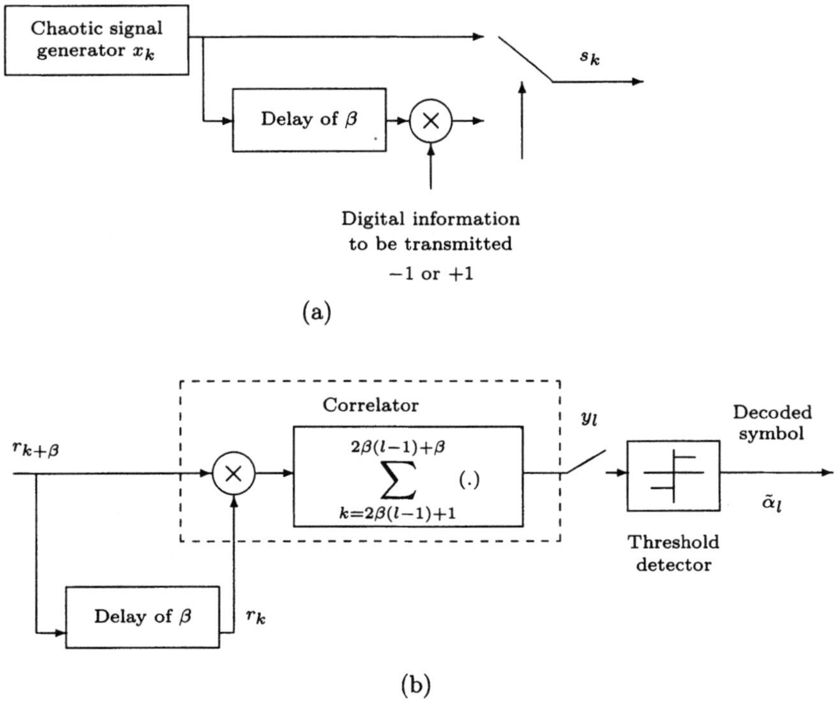

Fig. 8.4. Block diagram of a non-coherent DCSK system. (a) Transmitter; (b) receiver.

3B). We may now substitute (8.33) and (8.34) into (8.25) to obtain the BER, i.e.,

$$\mathrm{BER}_{\mathrm{CSK}} = \frac{1}{2}\mathrm{erfc}\left(\sqrt{\frac{2\beta}{1+4P_B+2N_0}}\right). \tag{8.35}$$

8.2.2 Non-Coherent DCSK System

In this section, we consider the non-coherent DCSK system. For this system, the basic modulation process involves dividing the bit period into two equal slots. The first slot carries a reference chaotic signal, and the second slot bears the information. For a binary system, the second slot is the same copy or an inverted copy of the first slot depending upon the symbol sent being "+1" or "−1". This structural arrangement allows the detection to be done in a non-coherent manner requiring no reproduction of the same chaotic carrying signals at the receiver. Essentially, the detection of a DCSK signal can be accomplished by correlating the first and the second slots of the same symbol and comparing the correlator output with a threshold. Figure 8.4 shows the block diagrams of a DCSK transmitter and receiver pair.

As in the CSK case, we assume that the DCSK signal is interfered by the DS-SS signal and corrupted by additive white Gaussian noise. Using the same notations as defined in Sect. 8.2.1, during the lth bit duration of the DCSK system, the transmitted DCSK signal can be written as

$$s_k = \begin{cases} x_k & \text{for } k = 2\beta(l-1)+1, 2\beta(l-1)+2, \ldots, 2\beta(l-1)+\beta \\ \alpha_l x_{k-\beta} & \text{for } k = 2\beta(l-1)+\beta+1, 2\beta(l-1)+\beta+2, \ldots, 2\beta l \end{cases} \tag{8.36}$$

whereas the kth transmitted signal for the DS-SS system is given by

$$u_k = \sqrt{P_B} b_k. \tag{8.37}$$

The received noisy signal, r_k, is given by

$$r_k = s_k + u_k + \xi_k \tag{8.38}$$

where ξ_k is as defined previously in Sect. 8.2.1.

At the DCSK receiver, the detector essentially calculates the correlation of the corrupted reference and data slots of the same symbol. We consider the output of the correlator for the lth received bit, y_l, which is given by

$$\begin{aligned} y_l &= \sum_{k=2\beta(l-1)+1}^{2\beta(l-1)+\beta} r_k r_{k+\beta} \\ &= \sum_{k=2\beta(l-1)+1}^{2\beta(l-1)+\beta} \left[x_k + u_k + \xi_k\right]\left[\alpha_l x_k + u_{k+\beta} + \xi_{k+\beta}\right] \\ &= \sum_{k=2\beta(l-1)+1}^{2\beta(l-1)+\beta} \Big[\alpha_l x_k^2 + \sqrt{P_B} b_{k+\beta} x_k + \alpha_l \sqrt{P_B} b_k x_k + P_B b_k b_{k+\beta} \\ &\qquad + \xi_{k+\beta} x_k + \alpha_l \xi_k x_k + \sqrt{P_B} \xi_{k+\beta} b_k + \sqrt{P_B} \xi_k b_{k+\beta} + \xi_k \xi_{k+\beta}\Big] \\ &= \underbrace{\alpha_l D}_{\text{required signal}} + \underbrace{\sqrt{P_B} F + \alpha_l \sqrt{P_B} G + P_B H}_{\text{interfering DS-SS signal}} \\ &\qquad + \underbrace{J + \alpha_l K + \sqrt{P_B} L + \sqrt{P_B} M + N}_{\text{noise}} \end{aligned} \tag{8.39}$$

where

$$D = \sum_{k=2\beta(l-1)+1}^{2\beta(l-1)+\beta} x_k^2 \tag{8.40}$$

$$F = \sum_{k=2\beta(l-1)+1}^{2\beta(l-1)+\beta} b_{k+\beta} x_k \tag{8.41}$$

$$G = \sum_{k=2\beta(l-1)+1}^{2\beta(l-1)+\beta} b_k x_k \tag{8.42}$$

$$H = \sum_{k=2\beta(l-1)+1}^{2\beta(l-1)+\beta} b_k b_{k+\beta} \tag{8.43}$$

$$J = \sum_{k=2\beta(l-1)+1}^{2\beta(l-1)+\beta} \xi_{k+\beta} x_k \tag{8.44}$$

$$K = \sum_{k=2\beta(l-1)+1}^{2\beta(l-1)+\beta} \xi_k x_k \tag{8.45}$$

$$L = \sum_{k=2\beta(l-1)+1}^{2\beta(l-1)+\beta} \xi_{k+\beta} b_k \tag{8.46}$$

$$M = \sum_{k=2\beta(l-1)+1}^{2\beta(l-1)+\beta} \xi_k b_{k+\beta} \tag{8.47}$$

$$N = \sum_{k=2\beta(l-1)+1}^{2\beta(l-1)+\beta} \xi_k \xi_{k+\beta}. \tag{8.48}$$

Suppose a "+1" is transmitted by the DCSK system for the lth symbol duration, i.e., $\alpha_l = +1$. Then, (8.39) can be re-written as

$$y_l \mid (\alpha_l = +1) = D + \sqrt{P_B}F + \sqrt{P_B}G + P_B H + J + K + \sqrt{P_B}L + \sqrt{P_B}M + N. \tag{8.49}$$

Likewise, given "−1" is sent during the lth symbol duration in the DCSK system, i.e., $\alpha_l = -1$, y_l becomes

$$\begin{aligned} y_l \mid (\alpha_l = -1) &= -D + \sqrt{P_B}F - \sqrt{P_B}G + P_B H + J - K \\ &\quad + \sqrt{P_B}L + \sqrt{P_B}M + N. \end{aligned} \tag{8.50}$$

The means and variances of the variables D to N can be shown equal to

$$\begin{array}{ll} \mathrm{E}[D] = \beta E[x_k^2] \equiv \beta P_s & \mathrm{var}[D] = \beta \mathrm{var}[x_k^2] \equiv \beta \Lambda \\ \mathrm{E}[F] = 0 & \mathrm{var}[F] = \beta P_s \\ \mathrm{E}[G] = 0 & \mathrm{var}[G] = \beta P_s \\ \mathrm{E}[H] = 0 & \mathrm{var}[H] = \beta \\ \mathrm{E}[J] = 0 & \mathrm{var}[J] = \beta P_s N_0/2 \\ \mathrm{E}[K] = 0 & \mathrm{var}[K] = \beta P_s N_0/2 \\ \mathrm{E}[L] = 0 & \mathrm{var}[L] = \beta N_0/2 \\ \mathrm{E}[M] = 0 & \mathrm{var}[M] = \beta N_0/2 \\ \mathrm{E}[N] = 0 & \mathrm{var}[N] = \beta N_0^2/4. \end{array} \tag{8.51}$$

where in the derivation of var$[D]$, it has been assumed that the autovariance of $\{x_k^2\}$ vanishes. Further, it can be readily shown that

$$\mathrm{cov}[\mu, \nu] = 0 \quad \forall\, \mu, \nu \in \{D, F, G, H, J, K, L, M, N : \mu \neq \nu\}. \tag{8.52}$$

Thus, the means and variances of $y_l \mid (\alpha_l = +1)$ and $y_l \mid (\alpha_l = -1)$ are given by

$$\mathrm{E}[y_l \mid (\alpha_l = +1)] = \beta P_s \tag{8.53}$$

$$\mathrm{E}[y_l \mid (\alpha_l = -1)] = -\beta P_s \tag{8.54}$$

$$\begin{aligned} \mathrm{var}[y_l \mid (\alpha_l = +1)] &= \mathrm{var}[y_l \mid (\alpha_l = -1)] \\ &= \mathrm{var}[D] + P_B \mathrm{var}[F] + P_B \mathrm{var}[G] \\ &\quad + P_B^2 \mathrm{var}[H] + \mathrm{var}[J] + \mathrm{var}[K] \\ &\quad + P_B \mathrm{var}[L] + P_B \mathrm{var}[M] + \mathrm{var}[N] \\ &= \beta \Lambda + 2\beta P_B P_s + \beta P_B^2 + \beta P_s N_0 \\ &\quad + \beta P_B N_0 + \frac{\beta N_0^2}{4}. \end{aligned} \tag{8.55}$$

As $y_l \mid (\alpha_l = +1)$ is the sum of a large number of random variables, we may assume that it is normally distributed when 2β is large. An error occurs when $y_l \leq 0 \mid (\alpha_l = +1)$, and the conditional probability is given by

$$\begin{aligned} \mathrm{Prob}(y_l \leq 0 \mid (\alpha_l = +1)) &= \frac{1}{2}\mathrm{erfc}\left(\frac{\mathrm{E}\left[y_l \mid (\alpha_l = +1)\right]}{\sqrt{2\mathrm{var}\left[y_l \mid (\alpha_l = +1)\right]}}\right) \\ = \frac{1}{2}\mathrm{erfc}&\left(\frac{\beta P_s}{\sqrt{2\beta\Lambda + 4\beta P_B P_s + 2\beta P_B^2 + 2\beta P_s N_0 + 2\beta P_B N_0 + \frac{\beta N_0^2}{2}}}\right). \end{aligned} \tag{8.56}$$

Likewise, given "−1" is sent, the conditional error probability is then equal to

$$\begin{aligned} \mathrm{Prob}(y_l > 0 \mid (\alpha_l = -1)) &= \frac{1}{2}\mathrm{erfc}\left(\frac{-\mathrm{E}\left[y_l \mid (\alpha_l = -1)\right]}{\sqrt{2\mathrm{var}\left[y_l \mid (\alpha_l = -1)\right]}}\right) \\ = \frac{1}{2}\mathrm{erfc}&\left(\frac{\beta P_s}{\sqrt{2\beta\Lambda + 4\beta P_B P_s + 2\beta P_B^2 + 2\beta P_s N_0 + 2\beta P_B N_0 + \frac{\beta N_0^2}{2}}}\right). \end{aligned} \tag{8.57}$$

Since all terms in (8.56) and (8.57) are independent of l, the bit error rate of the DCSK system under the interference of a DS-SS signal, denoted by $\mathrm{BER}_{\mathrm{DCSK}}$, equals the overall error probability of the lth transmitted symbol ($\mathrm{BER}_{\mathrm{DCSK}}^{(l)}$), i.e.,

$$\begin{aligned} \mathrm{BER}_{\mathrm{DCSK}} &= \mathrm{BER}_{\mathrm{DCSK}}^{(l)} \\ &= \mathrm{Prob}(\alpha_l = +1) \times \mathrm{Prob}(y_l \leq 0 \mid (\alpha_l = +1)) \\ &\quad + \mathrm{Prob}(\alpha_l = -1) \times \mathrm{Prob}(y_l > 0 \mid (\alpha_l = -1)) \\ &= \frac{1}{2}\mathrm{erfc}\left(\frac{\beta P_s}{\sqrt{2\beta\Lambda + 4\beta P_B P_s + 2\beta P_B^2 + 2\beta P_s N_0 + 2\beta P_B N_0 + \frac{\beta N_0^2}{2}}}\right) \end{aligned}$$

$$= \frac{1}{2}\text{erfc}\left[\left(\frac{2\Psi}{\beta} + \frac{4P_B}{\beta P_s} + \frac{2}{\beta}\left(\frac{P_B}{P_s}\right)^2 + \frac{2N_0}{\beta P_s} + \frac{2P_B N_0}{\beta P_s^2} + \frac{1}{2\beta}\left(\frac{N_0}{P_s}\right)^2\right)^{-\frac{1}{2}}\right] \quad (8.58)$$

$$= \frac{1}{2}\text{erfc}\left[\left(\frac{2\Psi}{\beta} + \frac{4P_B}{\beta P_s} + \frac{2}{\beta}\left(\frac{P_B}{P_s}\right)^2 + 4\left(\frac{E_b}{N_0}\right)^{-1} + \frac{4P_B}{P_s}\left(\frac{E_b}{N_0}\right)^{-1} + 2\beta\left(\frac{E_b}{N_0}\right)^{-2}\right)^{-\frac{1}{2}}\right] \quad (8.59)$$

where Ψ and E_b are defined as in (8.27) and (8.28). The expression given in (8.58) or (8.59) is then the analytical BER for the noisy DCSK signal under the influence of a DS-SS signal. Note that for fixed DS-SS signal power P_B and noise power spectral density $N_0/2$, the BER can be reduced by making similar adjustments as suggested in Sect. 8.2.1.

Moreover, when the DS-SS signal power is zero, i.e., $P_B = 0$, it can be readily shown that the bit error rate reduces to (see Sect. 5.2.2)

$$\begin{aligned} \text{BER}_{\text{DCSK}} \mid (P_B = 0) &= \frac{1}{2}\text{erfc}\left[\left(\frac{2\Psi}{\beta} + \frac{2N_0}{\beta P_s} + \frac{N_0^2}{2\beta P_s^2}\right)^{-\frac{1}{2}}\right] \\ &= \frac{1}{2}\text{erfc}\left[\left(\frac{2\Psi}{\beta} + 4\left(\frac{E_b}{N_0}\right)^{-1} + 2\beta\left(\frac{E_b}{N_0}\right)^{-2}\right)^{-\frac{1}{2}}\right]. \end{aligned} \quad (8.60)$$

Example — Consider the case where the logistic map is used for generating the chaotic sequence. We note that the autovariance of $\{x_k^2\}$ vanishes in this case (see Appendix 3B). By substituting (8.33) and (8.34) into (8.58), we obtain the BER of the DCSK system, i.e.,

$$\text{BER}_{\text{DCSK}} = \frac{1}{2}\text{erfc}\left(\sqrt{\frac{\beta}{1 + 8P_B^2 + 8P_B N_0 + 2N_0^2 + 8P_B + 4N_0}}\right). \quad (8.61)$$

8.3 Simulations and Evaluation

In this section the performance of the chaos-based digital communication systems under the influence of a DS-SS signal is studied by computer simulations. The logistic map described in Sect. 8.2.1 has been used to generate the chaotic sequences. In particular, the BER performance of the chaos-based digital communication systems will be investigated under variation of the following parameters:

- Average-bit-energy-to-noise-power-spectral-density ratio (E_b/N_0)
- Conventional-to-chaotic-signal-power ratio (P_B/P_s)
- Spreading factor (2β)

For comparison, we plot in each case also the analytical BERs obtained from the expressions derived in Sects. 8.2.1 and 8.2.2. The relevant simulated BERs for the CSK system are shown in Figs. 8.5 and 8.6, and those for the DCSK system are shown in Figs. 8.7 and 8.8. From these figures, we clearly see that the analytical and simulated BERs are in very good agreement, especially for large spreading factors (e.g., $2\beta = 100$ and 200) where the assumption of normal distribution of the conditional correlator output holds better. Also, as a general observation, the coherent CSK system consistently performs better than the non-coherent DCSK system under the influence of a coexisting conventional DS-SS signal. Further observations are summarized as follows.

- For the chaos-based communication systems under study, the BER generally decreases (improves) as the spreading factor or the average-bit-energy-to-noise-power-spectral-density (E_b/N_0) increases.
- The BERs increase (degrade) as P_B/P_s increases for a given E_b/N_0. This is apparently due to the increasing power of the DS-SS signal which causes more interference to the chaos-based systems, thus giving a higher BER.
- For the CSK system, at $E_b/N_0 = 7$ dB and a spreading factor of 100, the BER degrades from about 10^{-3} to 10^{-2} when the conventional-to-chaotic-signal-power ratio (P_B/P_s) increases from 0 dB to 10 dB. For the DCSK system, with the same increase in P_B/P_s, the BER now degrades from about 10^{-3} to 0.3 at $E_b/N_0 = 20$ dB. Thus, the CSK system is more tolerant to wideband interfering signals compared with the DCSK system.

In concluding our study of the coexistence problem, it is worth reiterating that faithful reproduction of the chaotic carriers at the receiver has been assumed for the coherent CSK system. Although this may not be a realistic assumption because of the technical difficulty associated with chaos synchronization (up to the time of writing), the results in this chapter provide a benchmark indicator for the performance of the coexisting systems. Nonetheless, as mentioned before, the problem is important technically since chaos-based systems are spread-spectrum systems which are expected to resist interfering and the kind of interference considered here, namely, one being generated from another conventional spread-spectrum system such as a DS-SS system, represents a realistic (future) practical concern when chaos-based systems need to "co-operate" with existing systems.

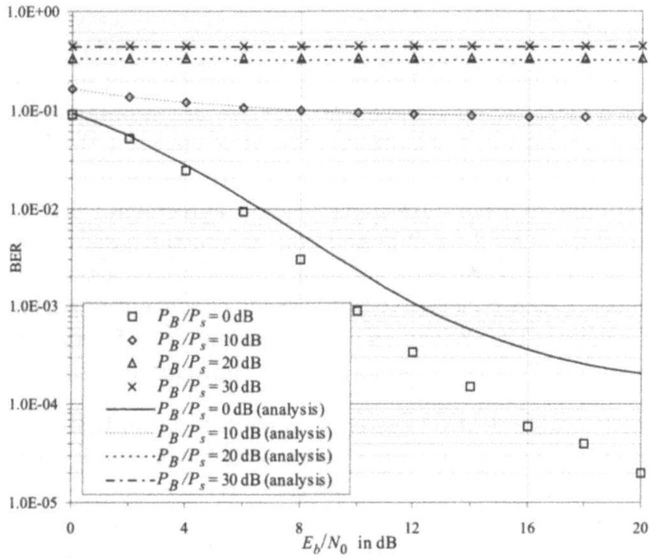

(a)

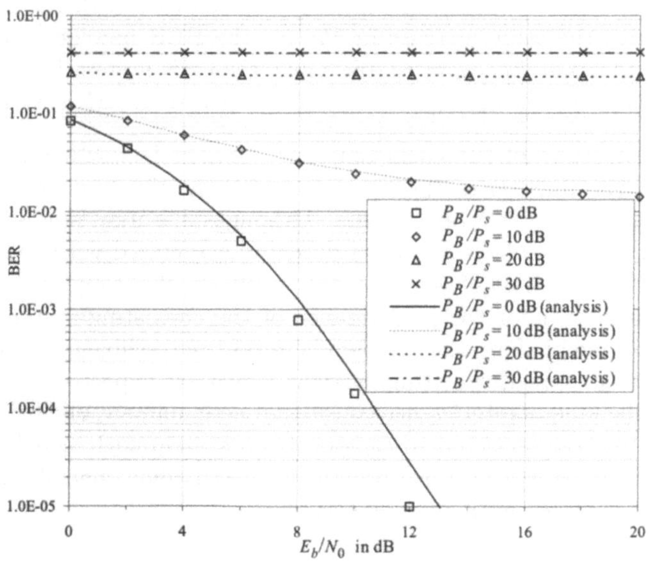

(b)

Fig. 8.5. BERs versus E_b/N_0 of the coherent CSK system under the interference of a direct-sequence-spread-spectrum signal. Simulated BERs are plotted as points and analytical BERs plotted as lines. (a) Spreading factor is 20; (b) spreading factor is 50.

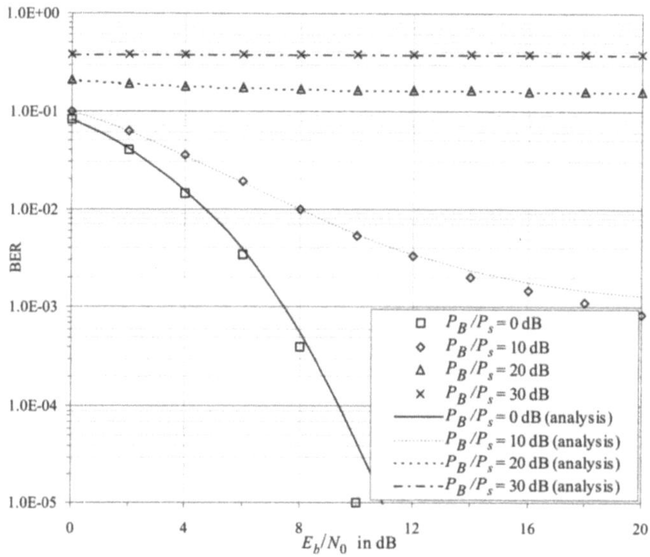

(a)

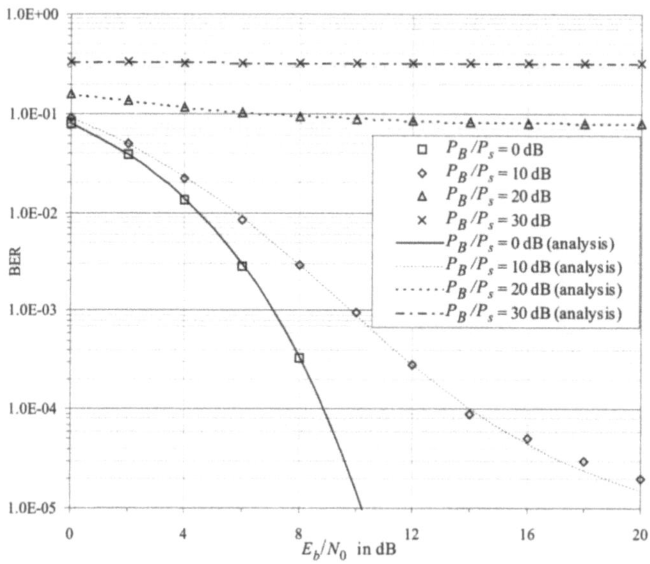

(b)

Fig. 8.6. BERs versus E_b/N_0 of the coherent CSK system under the interference of a direct-sequence-spread-spectrum signal. Simulated BERs are plotted as points and analytical BERs plotted as lines. (a) Spreading factor is 100; (b) spreading factor is 200.

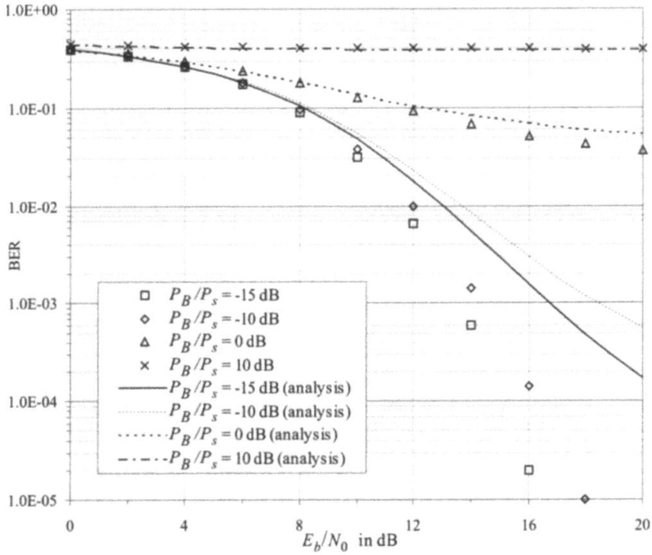

(a)

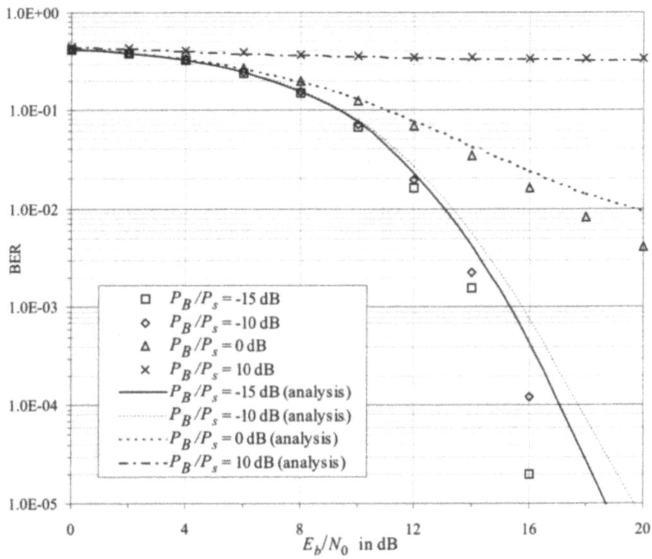

(b)

Fig. 8.7. BERs versus E_b/N_0 of the non-coherent DCSK system under the interference of a direct-sequence-spread-spectrum signal. Simulated BERs are plotted as points and analytical BERs plotted as lines. (a) Spreading factor is 20; (b) spreading factor is 50.

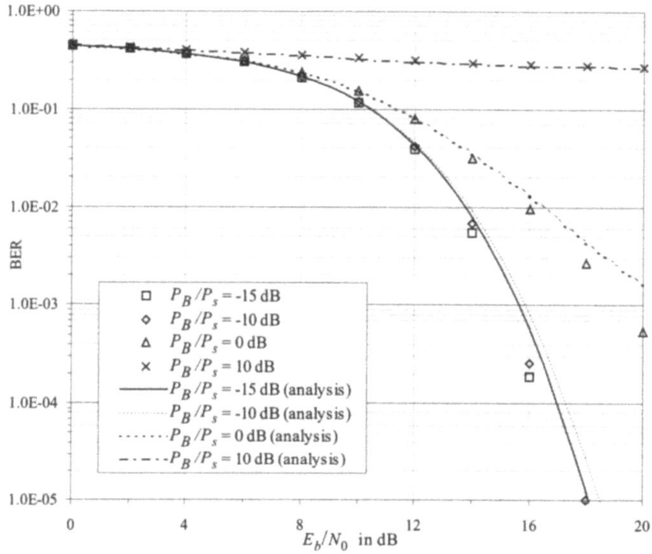

(a)

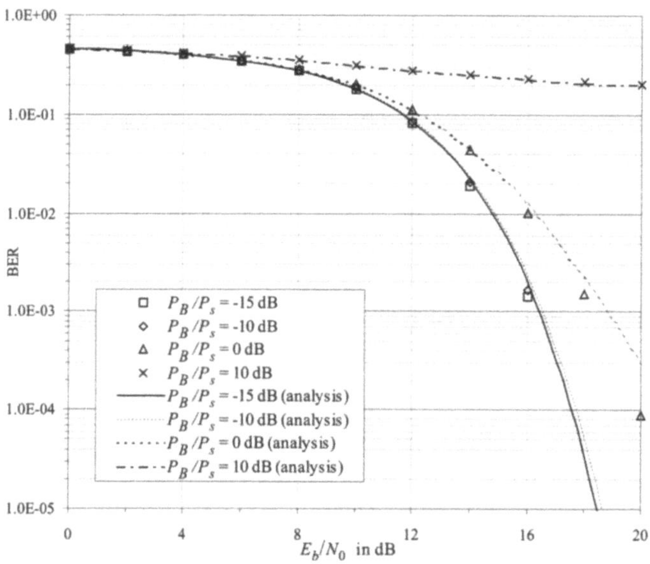

(b)

Fig. 8.8. BERs versus E_b/N_0 of the non-coherent DCSK system under the interference of a direct-sequence-spread-spectrum signal. Simulated BERs are plotted as points and analytical BERs plotted as lines. (a) Spreading factor is 100; (b) spreading factor is 200.

Appendix 8: Derivation of covariances and variances relevant to the analysis of CSK system

All symbols are as defined in Sect. 8.2.1.

$$
\begin{aligned}
\operatorname{cov}[A,B] &= \mathrm{E}[AB]-\mathrm{E}[A]\mathrm{E}[B] \\
&= \mathrm{E}\left[\left(\sum_{k=2\beta(l-1)+1}^{2\beta l} x_k^2\right)\left(\sum_{k=2\beta(l-1)+1}^{2\beta l} \sqrt{P_B}b_k x_k\right)\right] \\
&\quad - \mathrm{E}\left[\sum_{k=2\beta(l-1)+1}^{2\beta l} x_k^2\right]\mathrm{E}\left[\sum_{k=2\beta(l-1)+1}^{2\beta l} \sqrt{P_B}b_k x_k\right] \\
&= \sqrt{P_B}\mathrm{E}\left[\sum_{k=2\beta(l-1)+1}^{2\beta l} x_k^3 b_k + \sum_{\substack{k=2\beta(l-1)+1 \\ k\neq m}}^{2\beta l}\ \sum_{m=2\beta(l-1)+1}^{2\beta l} b_k x_k x_m^2\right] \\
&\quad - \sqrt{P_B}\left(\sum_{k=2\beta(l-1)+1}^{2\beta l} \mathrm{E}[x_k^2]\right)\left(\sum_{k=2\beta(l-1)+1}^{2\beta l} \mathrm{E}[b_k]\mathrm{E}[x_k]\right) \\
&= \sqrt{P_B}\sum_{k=2\beta(l-1)+1}^{2\beta l} \mathrm{E}[x_k^3]\mathrm{E}[b_k] \\
&\quad + \sqrt{P_B}\sum_{\substack{k=2\beta(l-1)+1 \\ k\neq m}}^{2\beta l}\ \sum_{m=2\beta(l-1)+1}^{2\beta l} \mathrm{E}[b_k]\mathrm{E}[x_k x_m^2] \\
&= 0
\end{aligned} \tag{8.62}
$$

$$
\begin{aligned}
\operatorname{cov}[A,C] &= \mathrm{E}[AC]-\mathrm{E}[A]\mathrm{E}[C] \\
&= \mathrm{E}\left[\left(\sum_{k=2\beta(l-1)+1}^{2\beta l} x_k^2\right)\left(\sum_{k=2\beta(l-1)+1}^{2\beta l} \xi_k x_k\right)\right] \\
&\quad - \mathrm{E}\left[\sum_{k=2\beta(l-1)+1}^{2\beta l} x_k^2\right]\mathrm{E}\left[\sum_{k=2\beta(l-1)+1}^{2\beta l} \xi_k x_k\right] \\
&= \mathrm{E}\left[\sum_{k=2\beta(l-1)+1}^{2\beta l} x_k^3 \xi_k + \sum_{\substack{k=2\beta(l-1)+1 \\ k\neq m}}^{2\beta l}\ \sum_{m=2\beta(l-1)+1}^{2\beta l} \xi_k x_k x_m^2\right] \\
&\quad - \left(\sum_{k=2\beta(l-1)+1}^{2\beta l} \mathrm{E}[x_k^2]\right)\left(\sum_{k=2\beta(l-1)+1}^{2\beta l} \mathrm{E}[\xi_k]\mathrm{E}[x_k]\right) \\
&= \sum_{k=2\beta(l-1)+1}^{2\beta l} \mathrm{E}[x_k^3]\mathrm{E}[\xi_k]
\end{aligned}
$$

$$
\begin{aligned}
&\quad + \sum_{\substack{k=2\beta(l-1)+1 \\ k\neq m}}^{2\beta l} \sum_{m=2\beta(l-1)+1}^{2\beta l} \mathrm{E}[\xi_k]\mathrm{E}[x_k x_m^2] \\
&= 0 \qquad (8.63) \\
\mathrm{cov}[B,C] &= \mathrm{E}[BC] - \mathrm{E}[B]\mathrm{E}[C] \\
&= \mathrm{E}\left[\left(\sqrt{P_B} \sum_{k=2\beta(l-1)+1}^{2\beta l} b_k x_k\right)\left(\sum_{k=2\beta(l-1)+1}^{2\beta l} \xi_k x_k\right)\right] \\
&\quad - \mathrm{E}\left[\sqrt{P_B} \sum_{k=2\beta(l-1)+1}^{2\beta l} b_k x_k\right] \mathrm{E}\left[\sum_{k=2\beta(l-1)+1}^{2\beta l} \xi_k x_k\right] \\
&= \sqrt{P_B}\mathrm{E}\left[\sum_{k=2\beta(l-1)+1}^{2\beta l} x_k^2 b_k \xi_k \right. \\
&\quad \left. + \sum_{\substack{k=2\beta(l-1)+1 \\ k\neq m}}^{2\beta l} \sum_{m=2\beta(l-1)+1}^{2\beta l} \xi_k x_k b_m x_m \right] \\
&\quad - \sqrt{P_B}\left(\sum_{k=2\beta(l-1)+1}^{2\beta l} \mathrm{E}[b_k]\mathrm{E}[x_k]\right)\left(\sum_{k=2\beta(l-1)+1}^{2\beta l} \mathrm{E}[\xi_k]\mathrm{E}[x_k]\right) \\
&= \sqrt{P_B}\left[\sum_{k=2\beta(l-1)+1}^{2\beta l} \mathrm{E}[x_k^2]\mathrm{E}[b_k]\mathrm{E}[\xi_k] \right. \\
&\quad \left. + \sum_{\substack{k=2\beta(l-1)+1 \\ k\neq m}}^{2\beta l} \sum_{m=2\beta(l-1)+1}^{2\beta l} \mathrm{E}[\xi_k]\mathrm{E}[x_k x_m]\mathrm{E}[b_m] \right] \\
&= 0 \qquad (8.64) \\
\mathrm{var}[A] &= \mathrm{var}\left[\sum_{k=2\beta(l-1)+1}^{2\beta l} x_k^2\right] \\
&= \sum_{k=2\beta(l-1)+1}^{2\beta l} \mathrm{var}[x_k^2] + \sum_{\substack{k=2\beta(l-1)+1 \\ k\neq m}}^{2\beta l} \sum_{m=2\beta(l-1)+1}^{2\beta l} \mathrm{cov}[x_k^2, x_m^2] \\
&= 2\beta \mathrm{var}[x_k^2] \qquad \text{apply (8.16)} \\
&= 2\beta\Lambda \qquad \text{where } \Lambda = \mathrm{var}[x_k^2] \qquad (8.65) \\
\mathrm{var}[B] &= \mathrm{var}\left[\sum_{k=2\beta(l-1)+1}^{2\beta l} \sqrt{P_B} b_k x_k\right]
\end{aligned}
$$

$$
\begin{aligned}
&= P_B \sum_{k=2\beta(l-1)+1}^{2\beta l} \mathrm{var}[b_k x_k] \\
&\quad + P_B \sum_{\substack{k=2\beta(l-1)+1 \\ k\neq m}}^{2\beta l} \sum_{m=2\beta(l-1)+1}^{2\beta l} \mathrm{cov}[b_k x_k, b_m x_m] \\
&= P_B \sum_{k=2\beta(l-1)+1}^{2\beta l} \left\{ \mathrm{E}[b_k^2 x_k^2] - \mathrm{E}^2[b_k x_k] \right\} \\
&\quad + P_B \sum_{\substack{k=2\beta(l-1)+1 \\ k\neq m}}^{2\beta l} \sum_{m=2\beta(l-1)+1}^{2\beta l} \{ \mathrm{E}[b_k x_k b_m x_m] \\
&\qquad - \mathrm{E}[b_k x_k]\mathrm{E}[b_m x_m] \} \\
&= P_B \sum_{k=2\beta(l-1)+1}^{2\beta l} \left\{ \mathrm{E}[b_k^2]\mathrm{E}[x_k^2] - \mathrm{E}^2[b_k]\mathrm{E}^2[x_k] \right\} \\
&\quad + P_B \sum_{\substack{k=2\beta(l-1)+1 \\ k\neq m}}^{2\beta l} \sum_{m=2\beta(l-1)+1}^{2\beta l} \{ \mathrm{E}[x_k x_m]\mathrm{E}[b_k b_m] \\
&\qquad - \mathrm{E}[x_k]\mathrm{E}[x_m]\mathrm{E}[b_k]\mathrm{E}[b_m] \} \\
&= 2\beta P_B P_s
\end{aligned}
\tag{8.66}
$$

$$
\begin{aligned}
\mathrm{var}[C] &= \mathrm{var}\left[\sum_{k=2\beta(l-1)+1}^{2\beta l} \xi_k x_k \right] \\
&= \sum_{k=2\beta(l-1)+1}^{2\beta l} \mathrm{var}[\xi_k x_k] \\
&\quad + \sum_{\substack{k=2\beta(l-1)+1 \\ k\neq m}}^{2\beta l} \sum_{m=2\beta(l-1)+1}^{2\beta l} \mathrm{cov}[\xi_k x_k, \xi_m x_m] \\
&= \sum_{k=2\beta(l-1)+1}^{2\beta l} \left\{ \mathrm{E}[\xi_k^2 x_k^2] - \mathrm{E}^2[\xi_k x_k] \right\} \\
&\quad + \sum_{\substack{k=2\beta(l-1)+1 \\ k\neq m}}^{2\beta l} \sum_{m=2\beta(l-1)+1}^{2\beta l} \{ \mathrm{E}[\xi_k x_k \xi_m x_m] - \mathrm{E}[\xi_k x_k]\mathrm{E}[\xi_m x_m] \} \\
&= \sum_{k=2\beta(l-1)+1}^{2\beta l} \left\{ \mathrm{E}[\xi_k^2]\mathrm{E}[x_k^2] - \mathrm{E}^2[\xi_k]\mathrm{E}^2[x_k] \right\} \\
&\quad + \sum_{\substack{k=2\beta(l-1)+1 \\ k\neq m}}^{2\beta l} \sum_{m=2\beta(l-1)+1}^{2\beta l} \{ \mathrm{E}[x_k x_m]\mathrm{E}[\xi_k \xi_m]
\end{aligned}
$$

$$
\begin{aligned}
& \quad - \mathrm{E}[x_k]\mathrm{E}[x_m]\mathrm{E}[\xi_k]\mathrm{E}[\xi_m]\} \\
&= 2\beta\left(\frac{N_0}{2}\right)P_s \\
&= \beta N_0 P_s
\end{aligned}
\tag{8.67}
$$

9. Techniques for Non-Coherent Detection in Chaos-Based Digital Communication Systems

Theoretically, coherent systems perform better than non-coherent systems, provided that the receiver can faithfully reproduce the same chaotic basis signals sent by the transmitter. However, the main technique that allows the reproduction of chaotic basis signals in the receiver, normally referred to as *chaos synchronization,* has yet to be proven feasible for low signal-to-noise conditions. Therefore, as the practicality of coherent systems is uncertain, researchers have considered non-coherent techniques for chaos-based communications [Kennedy and Kolumbán (2000a)]. The most widely studied non-coherent chaos-based digital communication system is the differential chaos-shift-keying (DCSK) system, which has been studied to some depth in the previous chapters. In this chapter, we review the basic requirements for non-coherent detection in chaos-based digital communication systems and describe a simple alternative approach for achieving the requirements.

9.1 Basic Requirements of Non-Coherent Detection

The basic feature of non-coherent communication systems is the omission of any synchronized replica of the basis signals in the receiver. This can be considered an advantage for chaos-based digital communications where technology has not yet permitted synchronized replica of the chaotic basis signals to be reliably generated in the receiver under noisy conditions. To develop non-coherent systems for communications, it is helpful to examine the basic requirements of non-coherent detection so that a variety of possible methods satisfying the requirements can be considered. From a simple theoretical viewpoint, a non-coherent digital communication system employs suitable modulation and demodulation processes that permit recovery of digital information without the presence of any synchronized replica of the chaotic basis signals in the receiver. In other words, the basic requirements are

- the modulation process in the transmitter should be designed to carry information with chaotic signals in a way that allows the receiver to recover the digital symbols from the noisy transmitted signal without reproducing the exact chaotic signals carrying the message; and

– the receiver should incorporate an effective demodulating scheme that minimizes the error probability given some or no knowledge about the chaotic carrying signals.

Referring to the familiar DCSK system [Kolumbán *et al.* (1996)], the design of the modulation process makes use of the bit structure to fabricate "distinguishable" digital symbols. Essentially, in this scheme, an information bit (digital symbol) is split into two halves. The first and second halves contain chaotic signals having a certain "detectable relationship". The receiver's job is to determine which digital symbol has been sent by identifying the kind of relationship between the first and second halves of the bit. In this system, however, the properties of the chaotic carrying signals need not to be used for demodulation.

In the chaos-shift-keying (CSK) system which is a generic form of modulation, one digital symbol is mapped to one chaotic basis signal. To design a non-coherent detector for this system, we need a procedure to distinguish the chaotic signals without the presence of the exact synchronized waveforms of the chaotic basis signals. In other words, we look for a distinguishable property of the chaotic signals that allow the detector to tell which digital symbol has been sent. The most commonly exploited distinguishable property has been the bit energy [Kolumbán *et al.* (1997a); Kennedy and Kolumbán (2000)]. However, when bit energy is chosen as the distinguishable property, detection can be accomplished easily by intruders, jeopardizing the security of the system.

Since chaotic signals are derived from deterministic dynamical systems (see Sect. 1.5.1), some deterministic properties of chaotic signals can be exploited for non-coherent detection. This area is not fully explored in the literature. In the rest of this chapter we describe a specific method in which some deterministic property of chaotic maps is used for non-coherent detection in a CSK system. Essentially, the method exploits the difference in the return maps of the chaotic signals representing the digital symbols. Two specific algorithms are discussed, namely, a *regression-based* algorithm and a *probability-based* algorithm.

9.2 Review of the CSK System

We consider a simple CSK system, in which the transmitter sends a signal consisting of chaotic signals extracted from two chaos generators. Essentially, the signal being transmitted toggles itself between the two sequences, $\{a_k\}$ and $\{b_k\}$, depending upon the value of the digital message to be sent, where k is the integer index for the sequence of values generated by the chaos generators, as shown in Fig. 9.1. In general the two chaotic sequences are generated by two chaotic maps, $f : S \to S$ and $g : S \to S$ for $S \subset \Re$, such that

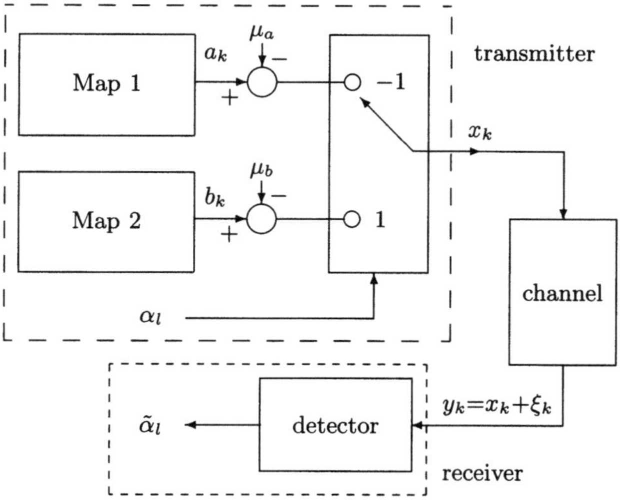

Fig. 9.1. Block diagram of the CSK system. α_l is the digital message and $\tilde{\alpha}_l$ is the restored message. μ_a and μ_b are the means of the chaotic sequences $\{a_k\}$ and $\{b_k\}$. ξ_k denotes white Gaussian noise.

$$\begin{aligned} a_{k+1} &= f(a_k) \\ b_{k+1} &= g(b_k). \end{aligned} \tag{9.1}$$

Now, consider the modulation process. Let α_l denote the lth bit, which is either "-1" or "$+1$". Also let 2β be the spreading factor, which is defined as the number of chaotic samples sent in one bit duration. The modulation proceeds as follows. In each bit duration, 2β consecutive values of either $\{a_k\}$ or $\{b_k\}$ are sent, depending upon the value of α_l. The output of the transmitter, x_k, during the lth bit duration, i.e., for $k = (l-1)2\beta + 1, (l-1)2\beta + 2, ..., 2l\beta$, is given by

$$x_k = \begin{cases} a_k - \mu_a, & \text{if } \alpha_l = -1 \\ b_k - \mu_b, & \text{if } \alpha_l = +1 \end{cases} \tag{9.2}$$

where μ_a and μ_b are the average values of the two chaotic sequences $\{a_k\}$ and $\{b_k\}$, respectively. Thus, the transmitter output, x_k, has a zero average. Further, assuming that the channel is subject to additive white Gaussian noise, the signal at the input to the receiver, y_k, is given by

$$y_k = x_k + \xi_k \tag{9.3}$$

where ξ_k is the added channel noise. At the receiving end, the general aim is to recover α_l with a minimum probability of error. Typical waveforms of transmitted signals are shown in Fig. 9.2.

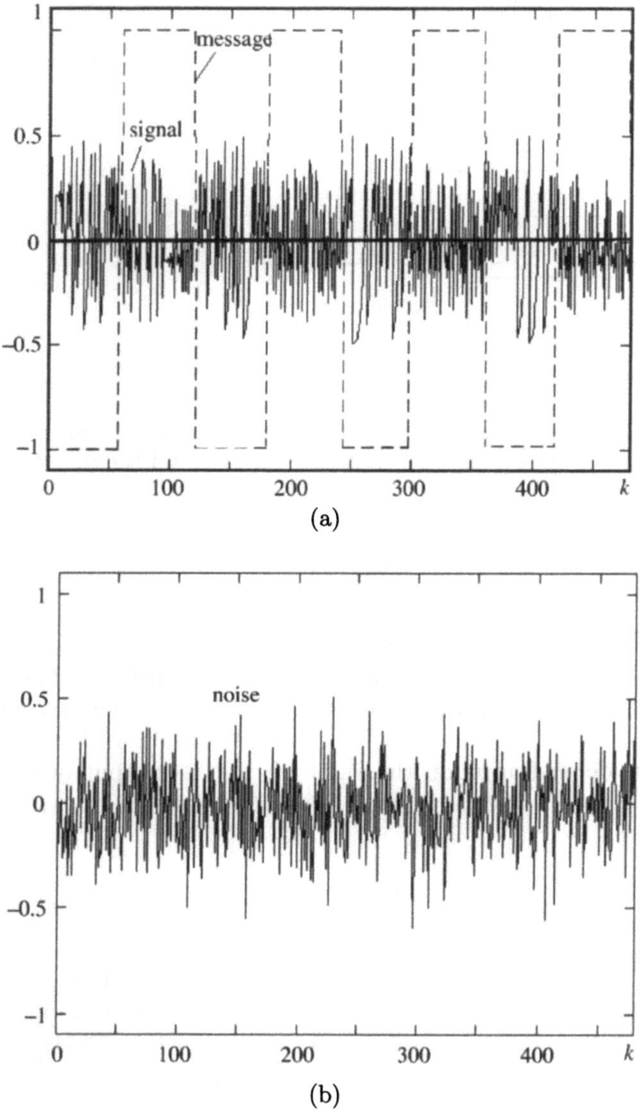

Fig. 9.2. Waveforms of (a) transmitted signal, and (b) channel noise at $E_b/N_0 = 15$ dB. Message is '$-1,+1,-1,+1,-1,+1,-1,+1,\ldots$' and spreading factor is 60. Signals are generated from the tent maps, i.e., $f(x) = 1 - 2|x - \frac{1}{2}|$ and $g(x) = 2|x - \frac{1}{2}|$.

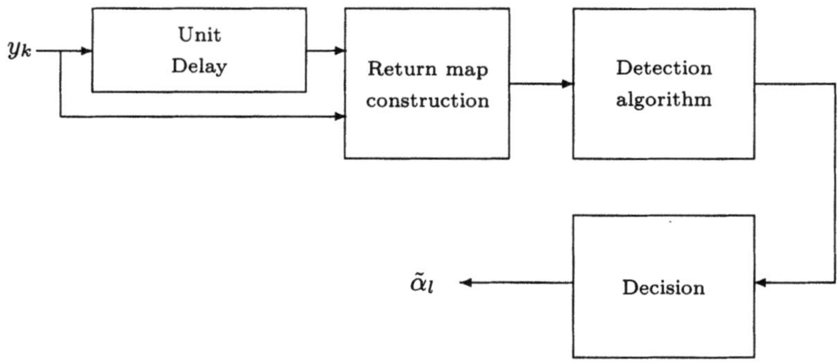

Fig. 9.3. Block diagram of the return-map-based detector.

9.3 Non-Coherent Detection Based on Return Maps

In this section we discuss specific detection approaches that make use of some deterministic property of the chaotic signals. Specifically, the return map will be exploited for detection, and two particular algorithms will be expounded. The first one is based on regression, and the second on maximizing the *a posteriori* probability. Figure 9.3 shows a block diagram of the return-map-based detector.

9.3.1 Regression Approach

As mentioned previously, the CSK system involves two chaotic maps. In the system under study, the chaotic maps are assumed to be written in a common form $h(.)$, with one distinguishing parameter p, i.e.,

$$\begin{aligned} \tilde{a}_{k+1} &= h(p, \tilde{a}_k) \\ \tilde{b}_{k+1} &= h(-p, \tilde{b}_k) \end{aligned} \tag{9.4}$$

where $(\tilde{a}_k, \tilde{a}_{k+1})$ and (a_k, a_{k+1}) are related by a simple transformation $T : S^2 \to S'^2$, and so are $(\tilde{b}_k, \tilde{b}_{k+1})$ and (b_k, b_{k+1}). That is,

$$(\tilde{a}_k, \tilde{a}_{k+1}) = T(a_k, a_{k+1}) \tag{9.5}$$

$$(\tilde{b}_k, \tilde{b}_{k+1}) = T(b_k, b_{k+1}). \tag{9.6}$$

As will become apparent, this requirement permits a simple regression procedure to be applied to the return map for effective detection. In particular, we consider the simple case where the parameter p appears explicitly as a multiplier to a common function $H(.)$, i.e.,

$$\begin{cases} \tilde{a}_{k+1} = pH(\tilde{a}_k) \\ \tilde{b}_{k+1} = -pH(\tilde{b}_k). \end{cases} \tag{9.7}$$

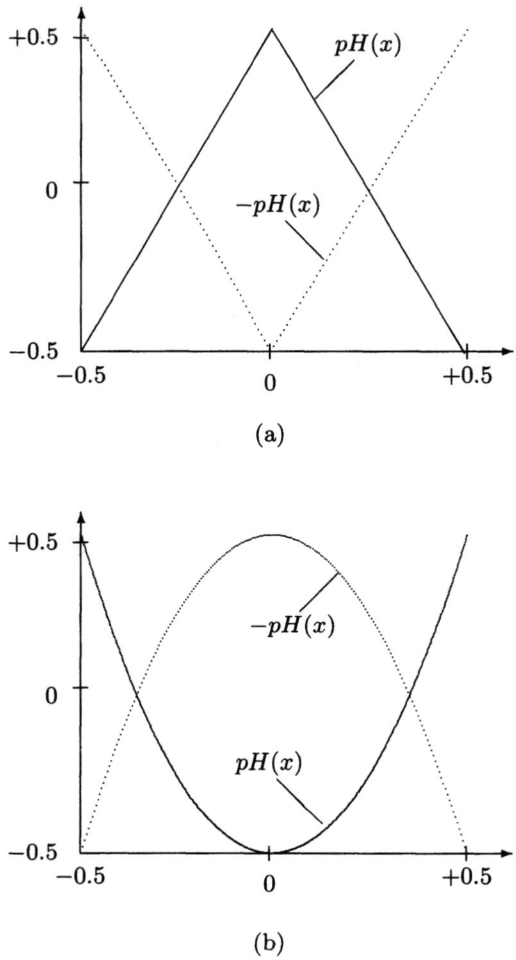

Fig. 9.4. Chaotic maps. (a) $H(x) = 0.5 - |2x|$, $p = 1$; (b) $H(x) = (x - \sqrt{2}/4)(x + \sqrt{2}/4)$, $p = 4$.

As an example, consider the chaotic maps employing the tent maps $f(x) = 1 - 2\left|x - \frac{1}{2}\right|$ and $g(x) = 2\left|x - \frac{1}{2}\right|$ for all $x \in S = (0, 1)$. The corresponding common form is given by

$$H(x) = \frac{1}{2} - |2x| \tag{9.8}$$

and $p = 1$. Here, the transformed domain and range are $S' = (-\frac{1}{2}, \frac{1}{2})$. See Fig. 9.4 (a).

Likewise, for the logistic maps $f(x) = 4x(1-x)$ and $g(x) = 1 - 4x(1-x)$, with $S = (0, 1)$, the corresponding common form, with $S' = (-\frac{1}{2}, \frac{1}{2})$, is

$$H(x) = \left(x - \frac{\sqrt{2}}{4}\right)\left(x + \frac{\sqrt{2}}{4}\right) \tag{9.9}$$

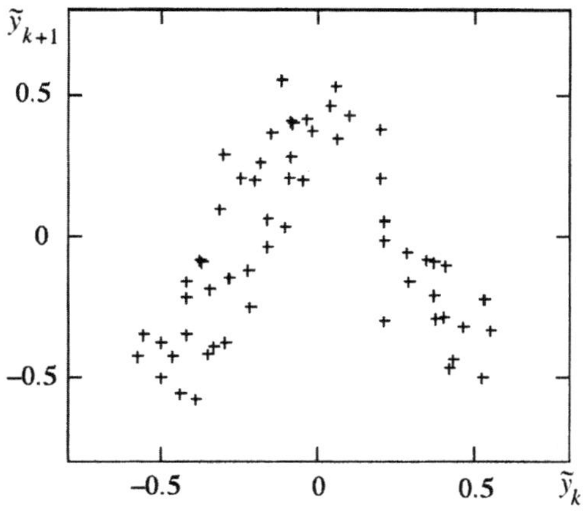

Fig. 9.5. Return map of received signal for symbol '−1' at $E_b/N_0 = 25$ dB. Spreading factor is 60.

and $p = 4$ in this case. See Fig. 9.4 (b).

The detection approach involves first collecting the points (y_k, y_{k+1}) for each bit. Then, the aforementioned T transformation is performed on these points, such that the resulting return map resembles either the curve $y = pH(x)$ or $y = -pH(x)$ under noise-free transmission, depending upon the transmitted bit α_l being "−1" or "+1". It should be noted that since the transmitter (as defined above) already removes the dc offset, the transformation may need to be adjusted to take into account the dc offset. For the two examples given above, no transformation is needed in the receiver because it is automatically done with the removal of the dc offset.

Under noisy condition, the points on the collected and transformed return map appear scattered, but to a certain extent (depending upon the noise level) remain close to the curves $y = \pm pH(x)$. A typical reconstructed return map is shown in Fig 9.5. The detection is formulated on the basis of a regression algorithm which aims to find the best fit of the curve

$$y = q_l H(x) \tag{9.10}$$

to the points of the return map for the lth bit. The regression here is to estimate the parameter q_l such that the set of points is closest to the above curve in a least-square sense (see Appendix 9). Once q_l is found, the decision rule can be as simple as

$$\tilde{\alpha}_l = \begin{cases} -1 & \text{if } q_l > 0 \\ +1 & \text{otherwise.} \end{cases} \tag{9.11}$$

The essence of the transformation T is to allow the two chaotic maps to be written with *only one* parameter which "strongly" characterizes the map. If the map is written in terms of two or more parameters, then these parameters will jointly characterize the map. Thus, the estimated value of any one parameter may not provide sufficient characterization of the particular chaotic map in order to allow accurate decision to be made as to which map has been sent. Hence, with only one parameter characterizing the map, the detection can be more accurately done.

9.3.2 Probability Approach

In this section we present an alternative approach for detection. The basis is still the return map, but the algorithm is based on maximizing the *a posteriori* probability.

Without loss of generality, we consider the received signal for the first symbol, i.e., $l = 1$. For convenience of the subsequent description, we denote the received signal block for the first bit by $\boldsymbol{y} = (y_1\ y_2\ \cdots\ y_{2\beta})$, and define the observation vectors as $\boldsymbol{v}_k = (y_k\ y_{k+1})$ for $k = 1, 2, \ldots, 2\beta - 1$. Define also $\boldsymbol{V} = (\boldsymbol{v}_1\ \boldsymbol{v}_2\ \cdots\ \boldsymbol{v}_{2\beta-1})$. Note that this is simply the return map constructed for the first bit. We decode the incoming chaotic signal block by selecting the symbol that would maximize the *a posteriori* probability given $\boldsymbol{V}$, i.e.,

$$\tilde{\alpha}_1 = \arg\max_{\alpha_1} \text{Prob}(\alpha_1 \text{ is sent} \mid \boldsymbol{V}). \tag{9.12}$$

As the *a posteriori* probability is not convenient to calculate, the Bayes' rule is applied to $\text{Prob}(\alpha_1 \text{ is sent} \mid \boldsymbol{V})$ to obtain

$$\text{Prob}(\alpha_1 \text{ is sent} \mid \boldsymbol{V}) = \frac{p(\boldsymbol{V} \mid \alpha_1 \text{ is sent})}{p(\boldsymbol{V})} \times \text{Prob}(\alpha_1 \text{ is sent}) \tag{9.13}$$

where $p(.)$ denotes the probability density function. Hence, (9.12) can be re-written as

$$\tilde{\alpha}_1 = \arg\max_{\alpha_1} p(\boldsymbol{V} \mid \alpha_1 \text{ is sent}) \tag{9.14}$$

because $\text{Prob}(-1 \text{ is sent}) = \text{Prob}(+1 \text{ is sent}) = 1/2$ and $p(\boldsymbol{V})$ is independent of α_1. In this detection scheme, we assume that $\boldsymbol{v}_k$ and $\boldsymbol{v}_m$ are independent for $k \neq m$. Thus, $p(\boldsymbol{V} \mid \alpha_1 \text{ is sent})$ in (9.14) can be expressed as

$$p(\boldsymbol{V} \mid \alpha_1 \text{ is sent}) = \prod_{k=1}^{2\beta-1} p(\boldsymbol{v}_k \mid \alpha_1 \text{ is sent}). \tag{9.15}$$

It should be pointed out that with this assumption, we effectively neglect the inter-dependence between the observation vectors, and thus the probability of an error occurring is expected to be larger than that of the optimal case studied by Hasler and Schimming [2000]. See also [Lau and Tse (2003)].

We also assume that in each of the observation vectors $\boldsymbol{v}_k = (y_k\ y_{k+1})$ ($k = 1, 2, \ldots, 2\beta - 1$), the initial condition of the chaotic signal x_k that gives rise to y_k is randomly selected from the chaotic range of the map according to the natural invariant probability density of the chaotic map. Now, suppose a "-1" is sent, i.e., $\alpha_1 = -1$, and the corresponding iterative map is f. Hence, we have

$$
\begin{aligned}
& p(\boldsymbol{v}_k \mid -1 \text{ is sent}) \\
= & \int_{-\infty}^{\infty} p(\boldsymbol{v}_k \mid (-1 \text{ is sent}, x_k)) \rho_f(x_k) dx_k \\
= & \int_{-\infty}^{\infty} \frac{1}{2\pi\sigma_n^2} \exp\left(-\frac{(y_k - x_k)^2 + (y_{k+1} - f(x_k))^2}{2\sigma_n^2} \right) \rho_f(x_k) dx_k \\
= & \frac{1}{2\pi\sigma_n^2} \int_{-\infty}^{\infty} \rho_f(x) \exp\left(-\frac{(y_k - x)^2 + (y_{k+1} - f(x))^2}{2\sigma_n^2} \right) dx \qquad (9.16)
\end{aligned}
$$

where $\rho_f(x)$ and $\sigma_n^2 (= N_0/2)$ denote the natural invariant probability density of f and variance of noise (noise power), respectively. Similarly, when a "+1" is sent and the corresponding iterative map is g, it is readily shown that

$$
\begin{aligned}
& p(\boldsymbol{v}_k \mid +1 \text{ is sent}) \\
= & \frac{1}{2\pi\sigma_n^2} \int_{-\infty}^{\infty} \rho_g(x) \exp\left(-\frac{(y_k - x)^2 + (y_{k+1} - g(x))^2}{2\sigma_n^2} \right) dx \qquad (9.17)
\end{aligned}
$$

where $\rho_g(x)$ represents the probability density of g. Note that since the transmitted signal contains no dc bias, the maps f and g should be chosen or translated appropriately such that they consistently contain no dc offset. The decision rule is simply given by

$$
\tilde{\alpha}_1 = \begin{cases} -1 & \text{if } \prod_{k=1}^{2\beta-1} p(\boldsymbol{v}_k \mid -1 \text{ is sent}) > \prod_{k=1}^{2\beta-1} p(\boldsymbol{v}_k \mid +1 \text{ is sent}) \\ +1 & \text{otherwise.} \end{cases} \qquad (9.18)
$$

The same detection algorithm applies to all other bits.

9.4 Simulations and Evaluation

The performance of the aforementioned detection methods is evaluated by computer simulations. We use the simple tent maps (Fig. 9.4 (a)) for chaos generation, and we present the bit error rate of the system for a range of spreading factors. Results for the two cases, corresponding to the regression-based and probability-based algorithms, are presented separately.

9.4.1 Regression Approach

For the regression-based detection, Fig. 9.6 shows the plots of the bit error rate (BER) versus the usual E_b/N_0. For comparison we show the performance of the most widely studied non-coherent DCSK system with $2\beta = 10$ and 20 [Kolumbán *et al.* (1996); Abel *et al.* (1998)]. It is fairer to compare the DCSK system for $2\beta = 20$ with our case for $2\beta = 10$ since half of the chaotic samples in DCSK are used as reference, though the DCSK with $2\beta = 20$ doubles the bandwidth requirement [Kisel *et al.* (2001)].

Next, we study the effect of varying the spreading factor 2β. From the simulation results, we may conclude that for constant E_b/N_0, the BER reaches a minimum for a certain spreading factor, as shown in Fig. 9.7. This behaviour occurs typically in non-coherent detection and can be explained as follows. For small spreading factors, detection accuracy is poor due to insufficient data points. Thus, detection improves as the spreading factor increases. However, as the spreading factor increases, the noise level increases accordingly for constant E_b/N_0, causing deterioration of the performance. For small spreading factors, the advantage that can be gained from increasing data points is more significant than the corresponding deterioration due to increased noise level. However, for large spreading factors, the noise admitted becomes excessive, and the advantage that can be gained from increasing data points becomes insignificant, causing overall deterioration of the performance.

9.4.2 Probability Approach

The tent maps are again employed to generate the chaotic sequences. In Fig. 9.8, the BERs versus E_b/N_0 are shown. We observe that for low E_b/N_0 values, using smaller spreading factors (e.g., 4 or 8) gives slightly better BER, whereas larger spreading factors produce comparatively lower BER for higher E_b/N_0. In Fig. 9.9, we plot the BERs versus the spreading factor in log scale. As in the regression-based detection method, for constant E_b/N_0, the BER achieves a minimum for a certain spreading factor. In this example, the minimum BERs are obtained with a spreading factor of about 10 for $E_b/N_0 = 12$, 14 and 16 dB.

It should be noted that the afore-described detection methods represent only two particular possibilities of exploiting the return map features, and they may not represent the most effective algorithm for detection. Nonetheless, our purpose is to demonstrate detection possibilities by suitably exploiting the chaotic determinism.

In this chapter we describe non-coherent methods for demodulating CSK signals, exploiting the built-in determinism of chaotic signals. We demonstrate specifically two methods exploiting the return maps for recovering the digital message carried by a CSK signal. The algorithm involves either a simple regression process or probability calculation. It should be stressed that methods exploiting deterministic properties of chaotic signals for non-coherent detection are still not exhausted and we expect more research activities to be initiated in this direction leading to better solutions for chaos-based digital communication systems.

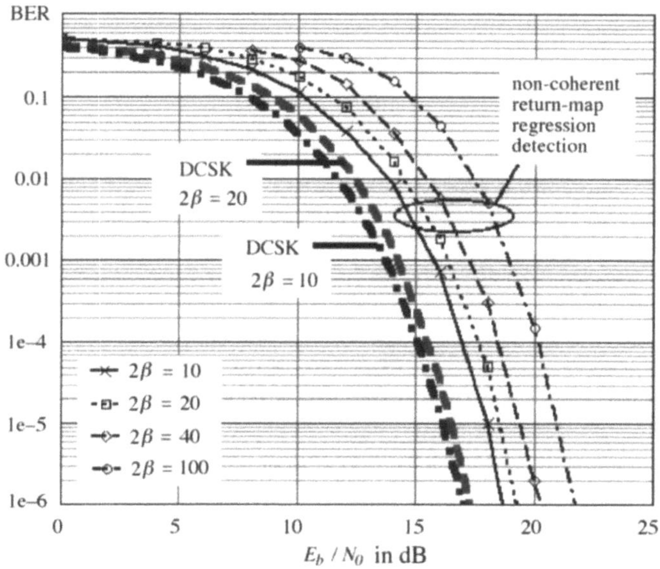

Fig. 9.6. BER versus E_b/N_0 using regression-based detection. Spreading factor $2\beta = 10$, 20, 40 and 100. Thick dash line corresponds to DCSK with $2\beta = 20$ which doubles the bandwidth requirement of our $2\beta = 10$ case, and thick dot-dash line corresponds to DCSK with $2\beta = 10$ which requires same bandwidth as our $2\beta = 10$ case.

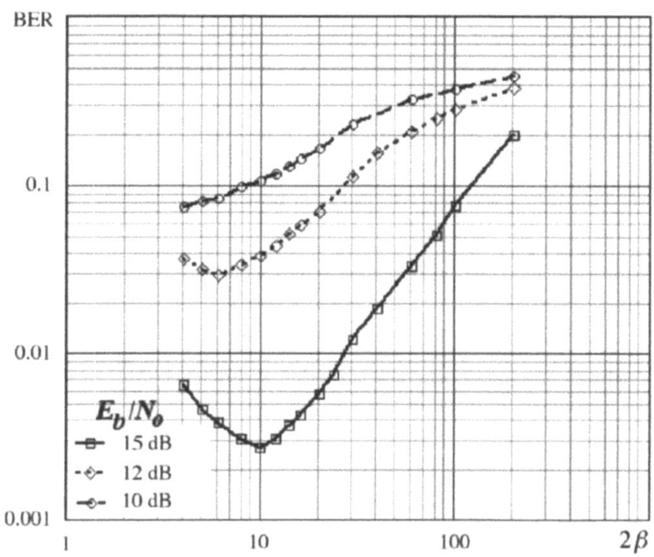

Fig. 9.7. Dependence of BER upon spreading factor 2β for the regression-based detector. The tent maps are used as chaos generators.

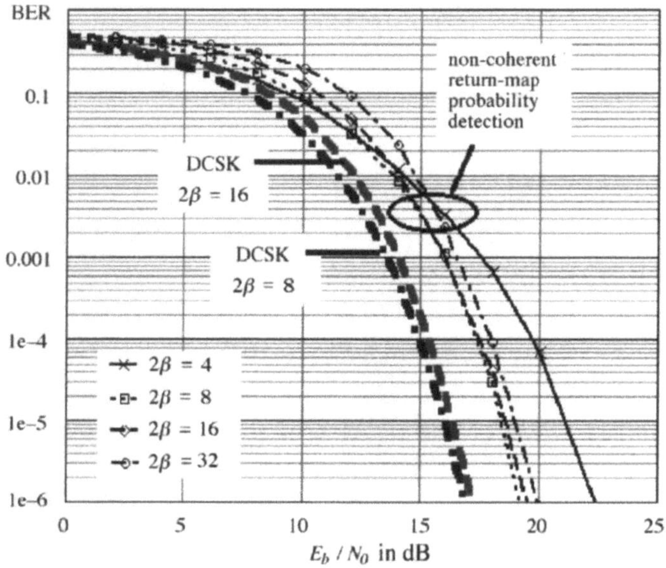

Fig. 9.8. BER versus E_b/N_0 using probability-based detection. Spreading factor $2\beta = 4$, 8, 16 and 32. The performance of the DCSK system is shown for comparison.

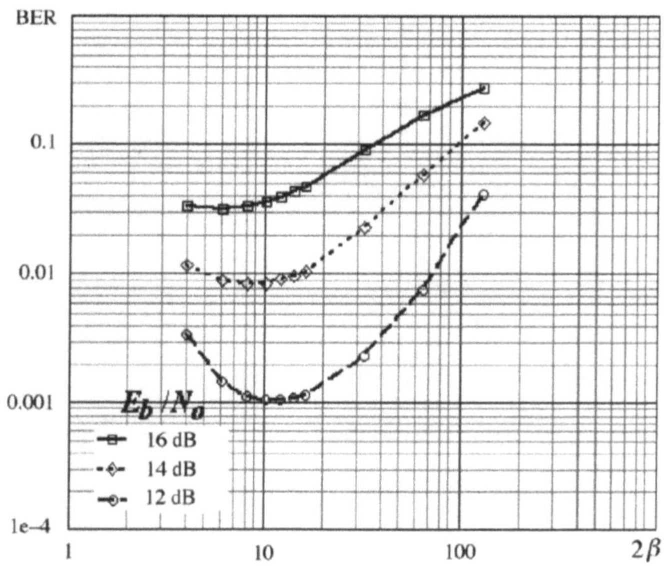

Fig. 9.9. Dependence of BER upon spreading factor 2β for the probability-based detector. The tent maps are used as chaos generators.

Appendix 9: Least squares estimate of parameter

The basic problem in the regression approach is to fit a curve of the form $y = qH(x)$ to a set of data points (x_1, y_1), (x_2, y_2), ..., (x_n, y_n). The fitting objective is to minimize the residual sum of squares, defined as

$$\mathrm{S} = \sum_{i=1}^{n} [y_i - qH(x_i)]^2 ,$$

where q is the parameter to be estimated. From elementary calculus, we set the partial derivative of the residual sum of squares with respect to q to zero [Freund and Wilson (1998)], i.e.,

$$\frac{\partial \mathrm{S}}{\partial q} = 0.$$

Expanding this, we get

$$\sum_{i=1}^{n} 2H(x_i) [y_i - (qH(x_i))] = 0.$$

The estimate of q is thus given by the following formula:

$$\hat{q} = \sum_{i=1}^{n} y_i H(x_i) \Big/ \sum_{i=1}^{n} H(x_i)^2.$$

References

Abel A., Schwarz W. (2002): Chaos communications — Principles, schemes, and system analysis, *Proceedings of the IEEE,* **90**(5), 691–710.

Abel A., Götz M., Schwarz W. (1998): Statistical analysis of chaotic communication schemes, *Proc. IEEE International Symposium on Circuits and Systems,* Monterey, USA, 465–468.

Abel A., Schwarz W., Götz M. (2000): Noise performance of chaotic communication systems, *IEEE Transactions on Circuits and Systems Part I,* **47**(12), 1726–1732.

Alligood K.T., Sauer T.D., Yorke J.A. (1996): *Chaos: An Introduction to Dynamical Systems,* Springer-Verlag, New York.

Anishchenko V.S., Pavlov A.N. (1998): Global reconstruction in application to multichannel communication, *Physical Review E,* **57**, 2455–2457.

Bray J., Sturman C.F., Mendolia J. (2001): *Bluetooth 1.1: Connect Without Cables,* Second Edition, Prentice Hall, Upper Saddle River.

Brzeźniak Z., Zastawniak T. (1998): *Basic Stochastic Processes: A Course Through Exercises,* Springer-Verlag, London.

Carroll T.L., Pecora L.M. (1999): Using multiple attractor chaotic systems for communication, *Chaos,* **9**, 445–451.

Chen G., Dong X. (1998): *From Chaos to Order: Methodologies, Perspectives and Applications,* World Scientific, Singapore.

Chen G., Ueta T., Editors (2002): *Chaos in Circuits and Systems,* World Scientific, Singapore.

Chen C.C., Yao K. (2000): Stochastic-calculus-based numerical evaluation and performance analysis of chaotic communication systems, *IEEE Transactions on Circuits and Systems Part I,* **47**(12), 1663–1672.

Clarke B. (1996): *GPS Aviation Applications,* McGraw-Hill, New York.

Cuomo K.M., Oppenheim A.V. (1993): Circuit implementation of synchronized chaos with applications to communications, *Physical Review Letters,* **71**, 65–68.

Dedieu H., Kennedy M.P., Hasler M. (1993): Chaos shift keying: modulation and demodulation of a chaotic carrier using self-synchronizing Chua's circuit, *IEEE Transactions on Circuits and Systems Part II,* **40**(10), 634–642.

Devaney R.L. (1992): *A First Course in Chaotic Dynamical Systems,* Addison-Wesley, New York.

Dixon R.C. (1994): *Spread Spectrum Systems with Commercial Applications,* Third Edition, John Wiley & Sons, New York.

Elbert B.R. (1999): *Introduction to Satellite Communication,* Artech House, Boston.

Feng J., Tse C.K. (2001): An on-line adaptive chaotic demodulator based on radial-basis-function neural networks, *Physical Review E,* **63**(2), 026202-1–10.

Feng J., Tse C.K., Lau F.C.M. (2002): Channel equalization for chaos-based communication systems, *IEICE Transactions on Fundamentals of Electronics, Communications and Computer Sciences,* **E85-A**(9), 2015–2024.

Freund R.J., Wilson W.J. (1998): *Regression Analysis: Statistical Modeling of a Response Variable,* Academic Press, San Diego.

Fu Y., Leung, H. (2001): Narrow-band interference cancellation in spread-spectrum communication systems using chaos , *IEEE Transactions on Circuits and Systems Part I,* **48**(7), 847–858.

Galias Z., Maggio G.M. (2001): Quadrature chaos shift keying, *Proc. IEEE International Symposium on Circuits and Systems,* Sydney, Australia, III313–III316.

Garg V.K. (1999): *IS-95 CDMA and cdma2000: Cellular/PCS Systems Implementation,* Prentice Hall, Upper Saddle River.

Gilhousen K.S., Jacobs I.M., Padovani R., Viterbi A.J., Weaver L.A. Jr., Wheatley C.E. III (1991): On the capacity of a cellular CDMA system, *IEEE Transactions on Vehicular Technology,* **40**(2), 303–312.

Hasler M., Schimming T. (2000): Chaos communication over noisy channels, *International Journal of Bifurcation and Chaos,* **10**, 719–735.

Hasler M., Schimming T. (2002): Optimal and suboptimal chaos receivers, *Proceedings of the IEEE,* **90**(5), 733–746.

Heidari-Bateni G., McGillem C.D. (1992): Chaotic sequences for spread spectrum: An alternative to PN-sequences, *Proc. IEEE International Conference on Selected Topics in Wireless Communications,* Vancouver, Canada, 437–440.

Heidari-Bateni G., McGillem C.D. (1994): A chaotic direct-sequence spread-spectrum communication system, *IEEE Transactions on Communications,* **42**(2/3/4), 1524–1527.

Hilborn R.C. (2000): *Chaos and Nonlinear Dynamics: An Introduction for Scientists and Engineers,* Oxford University Press, New York.

Hilliam R.M., Lawrance A.J. (2001): A statistical perspective on the dynamics of bivariate chaotic maps for communications modelling, *Proc. International Symposium on Nonlinear Theory and Its Applications,* Miyagi, Japan, 469–472.

Huang X., Leung H. (1995): Narrowband interference cancellation in a chaotic CDMA communication system, *Proc. IEEE Pacific Rim Conference on Communications, Computers, and Signal Processing,* Chiba, Japan, 437–440.

Itoh M., Murakami H. (1995): New communication systems via chaotic synchronizations and modulation, *IEICE Transactions on Fundamentals of Electronics, Communications and Computer Sciences,* **E78-A**(3), 285–290.

Jákó Z. (1998): Performance improvement of DCSK modulation, *Proc. International Workshop on Nonlinear Dynamics of Electronic Systems,* Budapest, Hungary, 119–122.

Jákó Z., Kis G., Kolumbán G. (2000): Multiple access capability of the FM-DCSK chaotic communications system, *Proc. International Workshop on Nonlinear Dynamics of Electronic Systems,* Catania, Italy, 52–55.

Kaplan D., Glass L. (1997): *Understanding Nonlinear Dynamics,* Springer-Verlag, New York.

Karim M.R., Sarraf M., Lawrence V.B. (2002): *W-CDMA and cdma2000 for 3G Mobile Networks,* McGraw-Hill, New York.

Kennedy M.P. (1998): Chaotic communications: from chaotic synchronization to FM-DCSK, *Proc. International Workshop on Nonlinear Dynamics of Electronic Systems,* Budapest, Hungary, 31–40.

Kennedy M.P., Kolumbán G. (2000): Digital communication using chaos, *Controlling Chaos and Bifurcation in Engineering Systems,* G. Chen (ed.), CRC Press, Boca Raton FL, 477–500.

Kennedy M.P., Kolumbán G., Guest Editors (2000a): Special Issue on Noncoherent Chaotic Communications, *IEEE Transactions on Circuits and Systems Part I,* **47**(12).

Kennedy M.P., Kolumbán G., Kis G., Jákó Z. (1998): Recent advances in communicating with chaos, *Proc. IEEE International Symposium on Circuits and Systems,* Montery, California, USA, 461–464.

Kennedy M.P., Kolumbán G., Kis G., Jákó Z. (2000): Performance evaluation of FM-DCSK modulation in multipath environments, *IEEE Transactions on Circuits and Systems Part I,* **47**(12), 1702–1711.

Kennedy M.P., Rovatti R., Setti G., Editors (2000a): *Chaotic Electronics in Telecommunications,* CRC Press, New York.

Kis G., Jákó Z., Kennedy M.P., Kolumbán G. (1998): Chaotic communications without synchronization, *Proc. 6th IEE Conference on Telecomunications,* Edinburgh, UK, 49–53.

Kisel A., Dedieu H., Schimming T. (2001): Maximum likelihood approaches for noncoherent communication with chaotic carriers, *IEEE Transactions on Circuits and Systems Part I,* **48**(5), 533–542.

Kocarev L., Halle K.S., Eckert K., Chua L.O., Parlitz U. (1992): Experimental demonstration of secure communications via chaotic synchronization, *International Journal of Bifurcation and Chaos,* **2**, 709–713.

Kohda T., Tsuneda A. (1994): Even- and odd-correlation functions of chaotic Chebyshev bit sequences for CDMA, *Proc. IEEE International Symposium on Spread Spectrum Technology and Applications,* Oulu, Finland, 391–395.

Kolumbán G. (1998): Performance evaluation of chaotic communications systems: Determination of low-pass equivalent model, *Proc. International Workshop on Nonlinear Dynamics of Electronic Systems,* Budapest, Hungary, 41–51.

Kolumbán G. (2000a): Basis function description of chaotic modulation schemes, *Proc. International Workshop on Nonlinear Dynamics of Electronic Systems,* Catania, Italy, 165–169.

Kolumbán G. (2000b): Theoretical noise performance of correlator-based chaotic communications schemes, *IEEE Transactions on Circuits and Systems Part I,* **47**(12), 1692–1701.

Kolumbán G. (2001): Frequency-modulated differential chaos shift keying (FM-DCSK): From theory to implementation, http://www.mit.bme.hu/research/chaos/inspect/index.html.

Kolumbán G., Kennedy M.P. (1998): The role of synchronization in digital communications using chaos — Part III: Performance bounds for correlation receivers, *IEEE Transactions on Circuits and Systems Part I,* **47**(12), 1673–1683.

Kolumbán G., Vizvari G.K., Schwarz W., Abel A. (1996): Differential chaos shift keying: a robust coding for chaos communication, *Proc. International Workshop on Nonlinear Dynamics of Electronic Systems,* Seville, Spain, 97–92.

Kolumbán G., Kennedy M.P., Chua L.O. (1997a): The role of synchronization in digital communication using chaos — Part I: Fundamentals of digital communications, *IEEE Transactions on Circuits and Systems Part I,* **44**(10), 927–936.

Kolumbán G., Kennedy M.P., Kis G. (1997b): Performance improvement of chaotic communications systems, *Proc. European Conference on Circuit Theory and Design,* Budapest, Hungary, 284–289.

Kolumbán G., Kennedy M.P., Kis G. (1997c): Multilevel differential chaos shift keying, *Proc. International Workshop on Nonlinear Dynamics of Electronic Systems,* Moscow, Russia, 191–196.

Kolumbán G., Kis G., Kennedy M.P., Jákó Z. (1997d): FM-DCSK: a new and robust solution to chaos communications, *Proc. International Symposium on Nonlinear Theory and Its Applications,* Hawaii, 117–120.

Kolumbán G., Kennedy M.P., Chua L.O. (1998a): The role of synchronization in digital communications using chaos — Part II: Chaotic modulation and chaotic synchronization, *IEEE Transactions on Circuits and Systems Part I,* **45**(11), 1129–1140.

Kolumbán G., Kis G., Jákó Z., Kennedy M.P. (1998b): A robust modulation scheme for chaotic communications, *IEICE Transactions on Fundamentals of Electronics, Communications and Computer Sciences,* **E81-A**(9), 1798–1802.

Kolumbán G., *et al.* (2002): The theory and implementation of a robust chaotic digital communications system, http://rfic.ucsd.edu/chaos/ws2002/Kolumban.pdf.

Kolumbán G., Kennedy M.P., Jákó Z., Kis G. (2002a): Chaotic communications with correlator receivers: theory and performance limits, *Proceedings of the IEEE,* **90**(5), 711–732.

Kraus J., Nossek J.A., Yang T., Chua, L.O. (2000): Evaluation of a continuous valued chaotic spreader used in a chaotic digital code-division multiple access ($(CD)^2MA$) system, *International Journal of Bifurcation and Chaos,* **10**, 1933–1950.

Lau F.C.M., Tse C.K. (2002): Optimum design for correlator-type receivers in chaos-based digital communications, *International Journal of Bifurcation and Chaos,* **12**, 1029–1038.

Lau F.C.M., Tse C.K. (2002a): Study of anti-jamming capabilities of chaotic digital communication systems, *Proc. International Symposium on Nonlinear Theory and Its Applications,* Xi'an, China, 65–68.

Lau F.C.M., Tse C.K. (2002b): Co-existence of chaos-based communication systems and conventional spread-spectrum communication systems, *Proc. International Symposium on Nonlinear Theory and Its Applications,* Xi'an, China, 107–110.

Lau F.C.M., Tse C.K. (2003): Approximate-optimal detector for chaos communication systems, *International Journal of Bifurcation and Chaos,* **13**.

Lau F.C.M., Tse C.K. (2003a): On optimal detection of noncoherent chaos-shift-keying signals in a noisy environment, *International Journal of Bifurcation and Chaos,* **13**.

Lau F.C.M., Yip M.M., Tse C.K., Hau S.F. (2002): A multiple access technique for differential chaos shift keying, *IEEE Transactions on Circuits and Systems Part I,* **49**(1), 96–104.

Lau F.C.M., Cheong K.Y., Tse C.K. (2002a): A permutation-based multiple access DCSK system, *Proc. International Symposium on Nonlinear Theory and Its Applications,* Xi'an, China, 511–514.

Lau F.C.M., Ye M., Tse C.K., Hau S.F. (2002b): Anti-jamming performance of chaotic digital communication systems, *IEEE Transactions on Circuits and Systems Part I,* **49**(10), 1486–1494.

Lau F.C.M., Cheong K.Y., Tse C.K. (2003): Permutation-based DCSK and multiple access DCSK systems, *IEEE Transactions on Circuits and Systems Part I,* **50**.

Lawrance A.J., Ohama G. (2002): Exact and analytically approximate bit error rates in chaos communication, *Proc. International Symposium on Nonlinear Theory and Its Applications,* Xi'an, China, 99–102.

Leung H., Lam J. (1997): Design of demodulator for the chaotic modulation and communication systems, *IEEE Transactions on Circuits and Systems Part I,* **44**(3), 262–267.

Leung H., Yu H., Murali K. (2002): Ergodic chaos based communication schemes, *Physical Review E,* **66**(3), 036203-1-8.

Liao T., Huang N. (1999): An observer-based approach for chaotic synchronization with applications to secure communication, *IEEE Transactions on Circuits and Systems I,* **46**(9), 1144–1150.

Maggio G.M., De Feo O. (2000): T-CSK: A robust approach to chaotic communications, *Proc. International Workshop on Nonlinear Dynamics of Electronics Systems,* Catania, Italy, 37–41.

Mazzini G., Setti G., Rovatti R. (1997): Chaotic complex spreading sequences for asynchronous DS-CDMA Part I: System modeling and results, *IEEE Transactions on Circuits and Systems Part I,* **44**(10), 937–947.

Mazzini G., Setti G., Rovatti R. (1998): Chaotic complex spreading sequences for asynchronous DS-CDMA Part II: Some theoretical performance bounds, *IEEE Transactions on Circuits and Systems Part I,* **45**(4), 496–506.

Milanović V., Zaghloul M.E. (1996): Improved masking algorithm for chaotic communication systems, *Electronics Letters,* **32**, 11–12.

Mullin T. (1993): *The Nature of Chaos,* Oxford University Press, New York.

Parlitz U., Chua L.O., Kocarev L., Halle K.S., Shang A. (1992): Transmission of digital signals by chaotic synchronization, *International Journal of Bifurcation and Chaos,* **2**, 973–977.

Peterson R.L., Ziemer R.E., Borth D.F. (1995): *Introduction to Spread Spectrum Communications,* Prentice Hall, Englewood Cliffs.

Proakis J.G. (1995): *Digital Communications,* McGraw-Hill, Singapore.

Proakis J.G., Salehi M. (1994): *Communications Systems Engineering,* Prentice Hall, Englewood Cliffs.

Ross S.M. (1993): *Introduction to Probability Models,* Fifth Edition, Academic Press, New York.

Rulkov N.F., Tsimring L.S. (1999): Synchronization methods for communications with chaos over band-limited channel, *International Journal of Circuit Theory and Applications,* **27**, 555–567.

Rulkov N.F., Sushchik M.M., Tsimring L.S., Volkovskii A.R. (2001): Digital communication using chaotic pulse position modulation, *IEEE Transactions on Circuits and Systems Part I,* **48**(12), 1436–1444.

Santamaria A., Lopez-Hernandez F.J., Editors (2001): *Wireless LAN Standards and Applications,* Artech House, Amsterdam.

Schimming T., Hasler M. (2000): Optimal detection of differential chaos shift keying, *IEEE Transactions on Circuits and Systems Part I,* **47**(12), 1712–1719.

Simon M.K., Omura J.K., Scholtz R.A., Levitt B.K. (2002): *Spread Spectrum Communications Handbook,* McGraw-Hill, New York.

Stewart G.W. (1998): *Matrix Algorithms,* Society for Industrial and Applied Mathematics, Philadelphia.

Strogatz S.H. (2001): *Nonlinear Dynamics and Chaos: With Applications to Physics, Biology, Chemistry and Engineering,* Perseus Books, New York.

Sushchik M., Tsimring L.S., Volkovskii A.R. (2000): Performance analysis of correlation-based communication schemes utilizing chaos, *IEEE Transactions on Circuits and Systems Part I,* **47**(12), 1684–1691.

Tam W.M., Lau F.C.M., Tse C.K., Hau S.F. (2001): Analytical BERs of coherent CSK and noncoherent DCSK systems with multiple users, *Proc. International Symposium on Nonlinear Theory and Its Applications,* Miyagi, Japan, 553–556.

Tam W.M., Lau F.C.M., Tse C.K., Yip M.M. (2002): An approach to calculating the bit error probability of a coherent chaos-shift-keying digital communication system under a noisy multi-user environment, *IEEE Transactions on Circuits and Systems Part I,* **49**(2), 210–223.

Tam W.M., Lau F.C.M., Tse C.K. (2002a): Multi-user detection techniques for multiple access chaos-based digital communication systems, *Proc. International Symposium on Nonlinear Theory and Its Applications,* Xi'an, China, 503–506.

Tam W.M., Lau F.C.M., Tse C.K. (2003): An approach to calculating the bit error rates of multiple access chaotic-sequence spread-spectrum communication systems employing multi-user detectors, *International Journal of Bifurcation and Chaos,* **13**.

Taylor H.M., Karlin S. (1998): *An Introduction to Stochastic Modeling,* Academic Press, New York.

Thompson J.M.T., Stewart H.B. (2002): *Nonlinear Dynamics and Chaos,* John Wiley & Sons, New York.

Torrieri D. (1986): The performance of five different metrics against pulsed jamming, *IEEE Transactions on Communications,* **34**(2), 200–204.

Tse C.K., Cheong K.Y., Lau F.C.M., Hau S.F. (2001): An approach for CSK detection based on return maps, *Proc. International Symposium on Nonlinear Theory and Its Applications,* Miyagi, Japan, 637–640.

Tse C.K., Lau F.C.M., Cheong K.Y., Hau S.F. (2002): Return-map based approaches for noncoherent detection in chaotic digital communications, *IEEE Transactions on Circuits and Systems Part I,* **49**(10), 1495–1499.

Verdu S. (1998): *Multiuser Detection,* Cambridge University Press, New York.

Viterbi A.J. (1995): *CDMA-Principles of Spread Spectrum Communication,* Addison-Wesley, Reading.

Volkovskii A.R., Tsimring L.S. (1999): Synchronization and communication using chaotic frequency modulation, *International Journal of Circuit Theory and Applications,* **27**, 569–576.

Wozencraft J.M., Jacobs, I.M. (1965): *Principles of Communication Engineering,* Wiley, New York.

Wu C.W., Chua L.O. (1993): Transmission of digital signals by chaotic synchronization, *International Journal of Bifurcation and Chaos,* **3**, 1619–1627.

Yang T., Chua L.O. (1997): Chaotic digital code-division multiple access communication systems, *International Journal of Bifurcation and Chaos,* **7**, 2789–2805.

Ziemer R.E., Tranter W.H. (1995): *Principles of Communications: Systems, Modulation and Noise,* Wiley, New York.

Index

MIX
Papier aus verantwortungsvollen Quellen
Paper from responsible sources
FSC® C105338
FSC
www.fsc.org

If you have any concerns about our products,
you can contact us on
ProductSafety@springernature.com

In case Publisher is established outside the EU,
the EU authorized representative is:
Springer Nature Customer Service Center GmbH
Europaplatz 3, 69115 Heidelberg, Germany

Printed by Libri Plureos GmbH
in Hamburg, Germany